Therapeutic Effects of Tea

Editors

B. Banerjee

T.C. Chaudhuri

Science Publishers, Inc.

Enfield (NH), USA   Plymouth, UK

CIP data will be provided on request.

SCIENCE PUBLISHERS, INC.
Post Office Box 699
Enfield, New Hampshire 03748
United States of America

Internet site: *http://www.scipub.net*

sales@scipub.net (marketing department)
editor@scipub.net (editorial department)
info@scipub.net (for all other enquiries)

ISBN 13: 978-1-57808-398-5 (hbk)

Published by Science Publishers, Inc., Enfield, NH, USA

Preface

All of us may have enjoyed a cup of tea at some time or the other, yet how many of us have delved into the fact that this is not merely a beverage, but nature's low calorie wonder drink – second only to water. Tea is the product of *Camellia sinensis*, is an unique non-alcoholic drink that cares, cures and of course cheers. It does fascinating things for the human body – keeps it glowing, beautiful, active and boosts the internal defense system. No other beverage can compare closely in so far as its multifunctional appeals, and effects are concerned. The quality and taste of tea are not one-dimensional; rather they appeal to a wide spectrum of people with varying tastes, quality preferences and health needs. Tea is the only beverage, which is both plebian and elitist in its appeal, from the royal to the common man.

What is in tea that unifies such disparate elements in the population? Is it legacy, tradition or simply effects of a hardcore publicity blitz? It is yet to be unravelled. But we do know that even in a fast changing world, where one is always looking for substitutes for foods and beverages with shifts in socioeconomic and lifestyle factors, tea still continues to occupy the centre-stage. From the whole cline of people having more time and less money, to those having more money but less time, tea continues to radiate its charm and romanticism.

The reasons for this optimism are not illusory but based on established facts. With the worldwide passion for foods and *drinks* that are natural in content, tea eminently fits in ways more than one. For nearly two centuries, tea is being grown in the countryside in an ambience that is totally ecological, apart from being a renewable natural resource. Even the processing is basically natural despite inevitable technical infringements that have only speeded up the process rather than changing the basic concept.

The added advantage of tea, is evident from the subsequent chapters, is its strong antioxidant activity. Although the therapeutic values of tea have been known for ages, current research, epidemiological studies and clinical evaluation have unequivocally established the superior abilities of tea compounds in preventing and curing a plethora of human diseases, some as dreadful as cancer. Interestingly enough, quality components of tea, coupled with its anti-oxidant properties, make tea a unique beverage.

Quality and therapeutic values, the two main attributes of tea, if projected in proper perspective, would appeal straight away to different segments of our population. Senior citizens comprising 20% of the population find the appeal of tea highly compelling because tea can be considered as an adjuvant therapy for age-related problems. Perhaps, it is the strong attraction for the aromatic stimulant in the brew which makes it the favourite among the health-conscious younger generation, and can be enjoyed in a variety of ways based on one's preference.

Our sincere gratitude to the reputed scientists for their valuable contributions which have been consolidated into this landmark research study. It is for the first time that data on medicinal properties of tea are presented, debated and discussed here by the scientists themselves with objective, rather than subjective observations.

A caveat must be entered here. It is not our intention to turn a cup of tea into a cup full of medicines by underpinning the therapeutic properties of tea! Rather, it is all the more significant to drink tea, as it is an all time rejuvenator and inexpensive solution for several ailments for people from all walks of life.

Cheers to excellent health with a hot cup of refreshing tea!

B. Banerjee
T.C. Chaudhuri

Contents

List of Contributors

Arora, Anu, Dr. MSc., PhD., Senior Research Fellow, Industrial Toxicological Research Centre, Mahatma Gandhi Road, Lucknow – 226001, U.P.

Banerjee, B., Dr. MSc., MS, PhD., SCD., FAS, Advisor, National Tea Research Foundation, Tea Board, 14 BTM, Sarani, Kolkata – 700001.

Banerjee, Shuvojit., Mr. MSC., Research Fellow, B.C. Guha Centre for Genetic Engineering and Biotechnology, University of Calcutta, 35 Ballygunge Circular Road, Kolkata – 700019.

Bhattacharyya, Arindam., Mr. MSc., Senior Research Fellow, Bose Institute, 93/1 A J C Bose Road, Kolkata – 700019.

Chatterjee, I.B., Prof. DSc., FNA, formerly Professor of Biochemistry and Currently Coordinator, B.C. Guha Centre for Genetic Engineering and Biotechnology, University of Calcutta, 35 Ballygunge Circular Road, Kolkata – 700019.

Chattopadhyay, D., Prof. PhD., FASC, FNASc, Professor of Biochemistry, and Coordinator, B.C. Guha Centre for Genetic Engineering and Biotechnology, 35 Ballygunge Circular Road, University of Calcutta, Kolkata – 700019.

Chaudhuri, T.C., Dr. MSc (Ag), PhD. Director (Research), Tea Board, 14 BTM Sarani, Kolkata – 700001.

Das, N.K., Mr. IAS, Chairman, Tea Board, 14 BTM Sarani, Kolkata – 700001.

Das, Tania, Dr. MSc., PhD., Scientist and Senior Lecturer, Animal Physiology Section, Bose Institute, 93/1, APC Bose Road, Kolkata – 700009.

Kalra, Neeta, Ms. MSc., Research Fellow, Industrial Toxicological Research Centre, Mahatma Gandhi Road, Lucknow – 226001, U.P.

Krishnan, R., Dr. MSc., PhD., Senior Research Fellow, Advanced Centre For Treatment, Research and Education in Cancer, Tata Memorial Centre, Navi Mumbai – 410208.

Lahiry, Laxmishri., Ms. MSc., Senior Research Fellow, Biology, Bose Institute, 93/1 APC Road, Kolkata – 700009.

Mandal, Debaprasad, Mr. MSc., Senior Research Fellow, Bose Institute, 93/1, APC Road, Kolkata – 700009.

Maru, G.B., Dr. MSc., PhD., Head, Tobacco Carcinogenesis, Advanced Centre for Treatment, Research and Education in Cancer (ACTREC), Tata Memorial Centre, Kharghar, Navi Mumbai – 410208.

(Late) Mishra, Anup, MSc., Research Fellow, B.C. Guha Centre for Genetic Engineering and Biotechnology, University of Calcutta, 35 Ballygunge Circular Road, Kolkata – 700019.

Mitra, Chandan, Prof. MSc., PhD., Sr. Professor and Head of Physiology Department, Presidency College, Kolkata – 700073.

(Late) Nag Chaudhuri, A.K., Prof. MSC., PhD., Director, School of Natural Product Studies, Division of Pharmacology, Dept. of Pharmaceutical Technology, Jadavpur University, Kolkata – 700032.

Pal, Sanjoy Kumar., Dr. PhD., Senior Scientist, Department of Gastroenterology, Sanjay Gandhi Postgraduate Institute of Medical Sciences, Raibareilly Road, Lucknow – 226014, U.P.

Patel, Rachana., Dr. MSc., PhD., Junior Research Fellow, Advanced Centre for Treatment, Research and Education in Cancer, Kharghar, Navi Mumbai – 410208.

Ramchandani, Asha, G., Dr. MSc., PhD., Senior Scientific Officer, ACTREC, Tata Memorial Centre, Kharghar, Navi Mumbai – 410208.

Ray, D.K., Mr. MSc., Research Fellow, Division of Pharmacology, Jadavpur University, Kolkata – 700032.

Sa, G., Dr. MSc., PhD., Reader and Senior Scientist, Animal Physiology Section, Bose Institute, 93/1, A P C Bose Road, Kolkata – 700009.

Sen, T., Dr. M Pharm, PhD., Lecturer in Pharmacology, Division of Pharmacology, Dept. of Pharmaceutical Technology, Jadavpur University, Kolkata – 700032.

Sharma, Archana, Prof. PhD., DSc. DSc(h), FNA, FASc, FNASc, INSA Emeritus Scientist, Centre for Advanced Study (Cell and Chromosome Research), Botany Department, University of Calcutta, 35, Ballygunge Circular Road, Kolkata – 700019.

Sharma, A.K., Prof. D.Sc., FNA, FASc, FNASc, FTWAS, National Professor, Centre for Advanced Study (Cell and Chromosome Research), Botany Department, University of Calcutta, 35 Ballygunge Circular Road, Kolkata – 700019.

Shukla, Yogeshwar, Dr. MSc., PhD., Sr. Assistant Director, Environmental Carcinogenesis Division, Industrial Toxicology Research Centre, Mahatma Gandhi Marg, Lucknow – 226001, U.P.

Talukdar, Geeta, Dr. MBBS, MD, PhD., FNASc, Senior Scientist and Consultant Genetics, Vivekananda Institute of Medical Sciences, 99, Sarat Bose Road, Kolkata – 700026.

Taneja, Pankaj, Dr. MSc., PhD., Post-Doctoral Fellow, John Hopkins NIEHS Centre for Human Health, Baltimore, Madison, USA.

Why Tea

A.K. Sharma
Botany Department,
University of Calcutta, India

Tea (*Camellia sinensis* L.O. Kuntze) is unique in more ways than one. It is both an agriculture and industry. It is unique in having a fairly large number of varieties in its natural populations. The Assam, the China and Cambod varieties are morphologically and anatomically distinct, along with several hybrids with many overlapping features. Some of these hybrids show strong characteristics of the Assam variety, while others have distinct China jat trait, the latter constituting the well-known China jat bushes common to the hills.

Despite the distinct morphological variations, diploid chromosome number is rather uniform in all varieties (2n = 30), though there are triploids, tetraploids, and even pentaploids. But hyper diploids are so far not found to be amenable for commercial exploitation. This mostly relates to their inability, as such, to produce an infusion of tea acceptable to the market. Their genetic potential is yet to be explored.

Tea is also unique being one of the few woody flowering perennials whose harvest is the leaf. To achieve optimum yield of the leaf, the plant's growth is manipulated and maintained as a bush in a state of continuous vegetative growth by periodic pruning. The degree of pruning determines the crop potential of tea and only the upper part of the bush produces the flush – the top two leaves and a bud which are plucked from the twigs on the plucking plateau. These are indeed novel features of tea and its cultivation.

Tea is also unique in its diverse medicinal properties. In nature, all species of plants, as the saying goes, have some medicinal properties, specific to particular diseases, but tea falls under that rare category of plants which are effective in preventing and resisting a variety of ailments. Tea compounds are effective against a wide array of human diseases ranging from gastric disorders to cancer, from osteoporosis to liver troubles. In addition to antistress effects, it is also claimed to possess anti-ageing properties. Researches are underway on the chemical and molecular basis of the action of tea constituents against various health disorders, their synthetic pathways and the site of synthesis.

It is well known that theaflavins and thearubigins in the processed tea leaves are the principal bioactive compounds of consequence to human health. They have complex chemical configurations and are the products of a series of biochemical reactions initiated with the picking of the famous two leaves and a bud. Green leaves have a rich polyphenol content, most important being the catechins. The thearubigins and theaflavins are the oxidation products of natural polyphenols, constituting a significant percentage of the black tea compounds on a dry weight basis.

As these two compounds per se would have such all pervading effects against so many human diseases, the question is often asked whether a cup of black tea infusion can provide some defense against all of them? If so, how?

Studies concentrating on individual diseases show the beneficial effects of tea extracts as against the infusions in the cup. Extracts containing catechin gallate directly act against or neutralize the disease causing agents at the cellular level. Epidemiological and clinical studies are progressing to ascertain their effect on human system *in vivo*.

Today India ranks No. 1 in the world production of tea, sharing 27% of the global production; but Indians drink tea the least, about 0.56 kg per capita per year. Is this low per capita adequate to avail the health benefits that tea offers? This is an issue which needs to be explored in detail and discussed in depth.

The present treatise encompasses all aspects of research and development of tea in relation to health — the state-of-art, emerging scenario and future strategy, covered by eminent experts in the field. I must congratulate the National Tea Research Foundation (NTRF) and particularly, Dr. Barun Banerjee, Advisor, NTRF, for bringing out such a comprehensive and extremely useful volume on Tea — the unique health drink.

FACT SHEET
TEA – THE KEY TO A HEALTHY LIFE
THERAPEUTIC VALUES

Source of tea plant	*Classification: Camellia sinensis*			
Where grown	Grown in about 34 countries (India, Kenya, African countries, Vietnam, Indonesia, China, Japan, Sri Lanka, Argentina, South America, etc.) However, finest teas are grown on tea estates or plantations ('gardens'). Quality is influenced by the location, climatic conditions, horticultural practices, etc.			
Basic Types of Tea (Though the leaves may be plucked from the same bush, it is only the difference in the method of processing that results in White, Green, Oolong or Black Tea)	**White Tea**	**Green Tea**	**Oolong Tea**	**Black Tea** (Regular, common or garden black tea)
Harvest	Buds or young leaves	Fresh, mature, green leaves	Fresh, mature, green leaves	Fresh, mature, green leaves
Manufacturing Process (Healthy difference is in the leaves)	Steamed (oxidase inactivation) (Not fermented)→ Dried	Withered→ Steamed or Panfried (oxidase inactivation)→ Dried	Withered→ Oxidised (briefly)→ Panfried (briefly)→ Dried	Withered→ Rolled→ Oxidised (fully)→ Fried→ Dried
Chemical Constituents of Tea	More than 700 chemical constituents are reported to be present in tea – Flavanols, Flavonoids (phytochemicals), Phenolic Acids, Carbohydrates, Vitamins, Minerals, Fluoride, Proteins, etc. (Polyphenols→Flavonols and Tannins, known as Catechins) (Catechins→Epigallocatechingallate [EGCG], Epicatechin-3-gllate, etc.) Theaflavins (TF)→Astringent compounds, which contribute to colour and taste of black tea brew. Thearubigins (TR) Nutrients (Carotenoids, Ascorbic Acid, etc.) Caffeine etc.			
Basic Qualities	Natural, Low-calorie, Non-alcoholic. Naturally rich in antioxidants			

Pharmacological Beneficial Effects on Human Health	–Power of antioxidants is correlated to Life Span Energy Potentials (LSEP). –Antioxidants and Vitamins prevent a number of diseases –Maintains homeostasis –Protects against DNA damage caused by Reactive Oxygen Species from endogenous and exogenous sources –Powerful chemopreventer of Reactive Oxygen and Nitrogen Species –Protective against oxidative damage in red blood cells –Positively influences bone mineral density (BMD) and protects against osteoporosis in elderly women –Inhibits growth and development of many types of Cancer (increases Apoptosis), atherosclerosis, etc. –Protective against Cancers induced by chemical carcinogens –Improves pathophysiological condition –Germicidal –Antimutagenic –Anticarcinogenic –Detoxifying function –Combats risk of developing heart disease, high blood cholesterol and high blood pressure. Acts as an Oxygen Radical Scavenger and exerts hypocholesterolemic action. –EGCG:- Treatment of EGCG to skin inhibits ultraviolet radiation induced oxidative stress. Protective effect against oxidative damage to cellular DNA –Significant increase of human plasma antioxidant capacity –Blocks production of Oxygen Free Radicals –Enhances the level of antioxidant defense enzymes such as SOD and catalase –Combats blood sugar (diabetes) –Protects cigarette smokers from CS-induced oxidative damage and the consequent degenerative diseases –Prevents dental caries, inhibits plaque formation –Helps build strong muscles –Aids growth, digestion and vitality –Helps better metabolic function –Prevents many illnesses that cause substantial mortality and morbidity –Prevents onset and subsequent progression of several diseases. Maintains body balance.

Chapter 1

Tea: A Global Perspective

B. Banerjee
National Tea Research Foundation, Kolkata, India

A. BLACK TEA

INTRODUCTION

From the perspective of health, or even otherwise, tea undoubtedly is the most widely consumed beverage (Trevisanto and Kim, 2001). The commercially traded tea by definition is derived from *Camellia sinensis* (L.) O. Kuntze, *C. assamica* spp *assamica* (Masters) Wight and *C. assamica* spp *lasioclayx* Wight or hybrids thereof (Wood and Barua, 1958). *Camellia sinensis* plant is a member of a evergreen family which grows in semi-arid tropics.

Even though all varieties of tea originate from the hybrids of the same species, i.e., *Camellia sinensis* (Banerjee, 1992), and therefore, of a common lineage, the resultant manufactured tea is generally classified as:

- Black Tea
- Green Tea
- Oolong Tea

depending on the degree of maceration of the leaf cells during the manufacturing process.

Difference between Black and Green Tea

It is only the variation in the method of processing that makes the colour of the tea leaves either black (regular tea) or green.

- *Black tea (commonly consumed with or without milk and sugar)*
  The process involves harvesting of green leaves, crushing them and exposing them to air. During this enzymatic process, the colour of the leaves change from green to brown, and subsequently after drying, turn into a beautiful rich deep-red brown-black colour, which gives it a delicious flavour.
- *Green Tea (normally consumed in Japan and China, without milk)*
  The fresh green leaves undergo least processing – they are quickly heated with steam and dried, so that they retain their green colour and unique aroma.

Both these varieties of tea are rich in antioxidants and help in better metabolic function.

However, based on worldwide consumption of tea, the largest share is attributable to black tea, accounting for 75 – 85%, followed by green tea (20 – 22%), and oolong tea (2 – 3%).

AN OVERVIEW

Black tea, because of the sheer magnitude of its consumption is, pivotal to the economics of many countries, producers and consumers alike. Tea is thus grown by some 34 countries spreading over Asia, Africa, South America and Australia (Table 1.1), within a fairly wide geographic swath, ranging from 27 S (Argentina) to 43 N (Georgia). The area under tea cultivation as well as its production and productivity in these countries vary, though it is difficult to relate the variation in productivity to any single factor or combination of factors (Table 1.2). The finest high quality tea is produced in the estates of Darjeeling.

Tea has strong adaptability growing as it does in both temperate and tropical climate. It needs an evenly distributed annual rainfall between 1150 and 5000 mm to grow on soils that are mostly kaolinitic. The conducive ambient temperature for good growth of tea varies from –8 C to 35 C depending on the area with daylight needs varying from 9 to 15 hrs. Given the wide agro-ecological conditions suitable for its growth along with its strong adaptability, tea truly is a ubiquitous plant (Banerjee, 2001a). Tea is also unique being both an agriculture and industry, and unlike other woody perennials its harvest is the leafy shoot – the two tender leaves and a bud.

Out of the 2.78 million hectares which are under tea cultivation globally (2002 data), India's share is 19% with a ranking of No. 2. Global production during the same period was 3099 million kg, India ranked No. 1 with a share

Table 1.1 Production of tea in different countries

(Quantity in '000 kgs)

Country	*1989*	*1990*	*1991*	*1999*	*2000*	*2001*	*2002*
India	688,105	720,338	754,192	825,935	846,922	853,923	826,165
Sri Lanka	207,951	234,074	241,552	284,149	306,794	296,301	310,032
Indonesia	146,790	160,481	152,898	161,003	157,371	161,202	166,027
Bangladesh	39076	45894	45030	45365	52639	57341	52863
China	534,900	540,100	541,600	675,871	683,324	701,699	715,000
Taiwan	22130	22299	21378	22555	20349	19837	20000
Japan	90501	89903	87903	88512	89309	89809	90000
Vietnam	30200	32200	33000	65000	70000	80000	84000
Papua New Guinea	6100	6200	5500	6200	6200	6100	6200
USSR/CIS	127,000	131,000	116,000	15930	16900	17500	18300
Iran	46000	44000	45000	60000	53000	46000	48000
Turkey	141,582	126,768	136,887	170,563	130,671	142,900	145,000
Kenya	180,600	197,008	203,589	248,818	236,286	294,631	287,044
Uganda	4620	6704	8950	24730	29282	33255	33831
Tanzania	16977	18414	19321	23490	23897	24745	27511
Malawi	39476	39059	40530	38469	42114	36770	39185
Mozambique	1800	2000	2000	1800	2500	3000	3000
Zaire	3100	3000	3000	2500	2500	2600	2700
Mauritius	5500	5751	5918	1463	1313	1517	1382
Rwanda	12971	12854	13546	12970	14391	17809	17000
Burundi	3862	4040	5320	6859	7118	9011	6605

Contd.

Table 1.1 Contd.

(Quantity in '000 kgs)

Country	*1989*	*1990*	*1991*	*1999*	*2000*	*2001*	*2002*
Cameroon	2407	2941	3351	4485	4004	4200	4200
Argentina	40000	43000	43700	55000	63000	59000	62000
Brazil	11000	9500	9500	7500	7000	7000	7200
Peru	3000	3500	3500	2400	2500	2600	2700
Ecuador	2000	1500	2036	1800	1700	1600	1700
Zimbabwe	17948	17092	15163	20411	22489	22382	22544
Myanmar	49799	36150	47300	60000	60000	60000	60000
Thailand	6115	7565	7600	8671	8700	8700	8700
Malaysia	5678	5621	6984	6246	5642	5413	5500
South Africa	12739	12233	12152	10570	10612	10734	10900
Australia	713	773	720	1300	1300	1300	1400
Ethiopia	1498	1500	1500	4200	4500	4600	4700
Nepal	1050	1200	1583	7200	7500	7700	7900
Grand total	*2,503,168*	*2,584,662*	*2,638,653*	*2,972,975*	*2,991,827*	*3,091,179*	*3,099,289*

Source: Tea Board

Table 1.2 Area under tea in different countries

(Area in hectares)

Country	*1989*	*1990*	*1991*	*1999*	*2000*	*2001*	*2002*
India	414,953	416,269	420,470	490,200	504,366	509,770	511,940
Sri Lanka	222,110	221,758	221,691	195,460	188,971	188,971	188,971
Indonesia	129,375	134,934	137,055	156,840	157,488	160,991	162,000
Bangladesh	47439	47650	47677	48611	49195	49313	49500
China	1,065,866	1,061,864	1,060,530	1,130,000	1,089,000	1,140,700	1,155,000
Taiwan	23914	24315	23863	20502	19701	18938	18950
Japan	59000	58500	57600	50700	50400	50100	50200
Vietnam	65000	59900	65000	77142	80000	82000	84000
Papua New Guinea	4000	4000	4000	4000	4000	3800	3850
USSR/CIS	81400	81400	81400	40600	40600	40600	40600
Iran	32000	32000	32000	34664	34664	34664	34664
Turkey	89525	90575	88639	76749	76749	76653	76700
Kenya	87473	97020	100,629	120,429	122,236	131,581	132,000
Uganda	20905	20905	20500	20400	20570	20570	20570
Tanzania	18875	18875	19415	21479	21212	21371	21400
Malawi	18507	18204	18300	18807	18782	18761	18780
Mozambique	2000	2000	2000	2000	2800	3200	3300
Zaire	10000	9500	9000	6000	6000	6000	6000
Mauritius	3071	2905	2869	671	670	659	659
Rwanda	11929	12286	12395	11707	12300	12825	12900
Burundi	6840	8537	7866	7862	8200	8709	9000
Cameroon	1486	1486	1486	1547	1546	1546	1546

Contd.

Table 1.2 Contd.

(Area in hectares)

Country	*1989*	*1990*	*1991*	*1999*	*2000*	*2001*	*2002*
Argentina	41276	41276	41406	39000	39000	36000	36000
Brazil	5000	5500	5500	5500	5500	5000	5000
Peru	4000	3000	3000	2500	2600	2700	2750
Ecuador	850	850	850	940	940	950	950
Zimbabwe	6102	6337	6352	6772	6824	6850	6875
Myanmar	58150	58662	57683	68392	68392	68392	68392
Thailand	6574	6066	6066	9412	9412	9412	9412
Malaysia	2666	2513	3056	3051	3100	3100	3100
South Africa	6721	6326	6721	6957	6818	7335	7500
Australia	658	775	775	800	800	800	800
Ethiopia	1619	1619	1619	2200	2300	2350	2400
Nepal	2200	2400	2400	10250	10250	10250	10250
Grand total	*2,551,484*	*2,560,207*	*2,569,813*	*2,692,144*	*2,665,386*	*2,734,861*	*2,755,959*

Source: Tea Board

of 27%. Global export of tea stood at 1431 million kgs during 2002 with India occupying the fourth position, with an export share of 14%.

THE HARVEST

Young shoots consisting mostly of top two fresh green leaves and a bud constitute the harvest of tea called fine plucking and this accounts for about 10-12% of the total biomass of tea. Hence, unlike other perennials, tea has a low harvest index which is difficult to surmount (Banerjee, 2001b). These shoots prized for their flavour also contain the compounds of pharmaceutical importance, are harvested at intervals of seven to ten days. To effect the production of these shoots, tea is maintained as a low bush in a state of continuous vegetative growth, by periodic productive pruning at pre-determined heights, usually at about 4 to 5 ft to facilitate plucking of the top leaves by hand.

To remain in a productive state, tea also needs an optimal nutritional balance generally provided by matching combinations of nitrogen, potash and phosphate, and micronutrients. Crop protection is a major component of tea husbandry, albeit there are restrictions on the use of pesticides and the statutory need for producing residue-free tea. Aspects of tea growing are detailed in the publications by Banerjee (2002, 2001a), Graham (1984) and Barua (1987).

PRODUCERS OF BLACK TEA

From the perspective of human health, black tea assumes great significance because 80% of the global tea drinkers have a strong preference for it. This is also reflected in the Indian tea statistics which indicate 77% of the Indian produce is black tea and 23% green tea. Generally, but not always, 59% of the Indian black tea is manufactured by the CTC (crushed, torn and curled) method and the balance by the orthodox method. In India, black tea is produced in Assam, North Bengal, Darjeeling, North East States and in the South.

Apart from India, major producers of CTC black tea include Kenya and other African countries, Vietnam and Indonesia. China is steadily increasing its production of black tea. Orthodox black tea producers include Sri Lanka, Vietnam and Indonesia in Asia, and Argentina in South America. However, Sri Lanka leads in tea export.

TEA PROCESSING IN RELATION TO RELEASE OF THERAPEUTIC COMPOUNDS

Black tea processing begins with the use of only freshly hand picked young two leaves and a bud because mature leaves lead to a loss in quality and perhaps pharmaceutical compounds as well.

The prime factor in tea processing is the conversion of polyphenols constituting about 30-40% of the chemicals in the tea leaf on dry weight basis into catechins. The genesis of both tea quality and the production of therapeutic compounds can be traced to the reactions involved in this conversion process. The major physical process involved both in the CTC and orthodox systems is the maceration of the leaves in varying degrees so that the catechins held in cell vacuoles get mixed with cytoplasmic polyphenol oxidase enzyme.

In the orthodox process, the degree of cell rupture during maceration (rolling) is relatively low, but it is more thorough in the CTC process, leading to highly effective efficient mixing of catechins with the polyphenol oxidase enzyme. This variation is significant because it not only establishes the distinct quality profile of CTC and orthodox varieties (Takeo and Mahanta, 1972), but the dynamics of the therapeutic compounds as well. This is universally true of the black tea produced worldwide.

Withering, the first step in processing ensures a reduction in the moisture level in the harvest by blowing hot air over the raw material for 8 to 18 hours in a special withering chamber. During this process, the moisture of the tea drops to between 60 (soft wither) and 50% (hard wither) thereby softening the leaves, to enable rolling. Though withering appears to be a physical process, it also involves complex interplays of chemicals. The most significant of these is the production of precursors of the compounds that determine the flavour of tea, and initiation of the production of pharmacological compounds (Wood and Chanda, 1955).

Preconditioning of the withered leaves (10-15 minutes) without any physical pressure, allows maximal utilization of polyphenol oxidase enzyme for optimal fermentation, which is the next crucial stage in tea processing. This enzyme, located in the cell epidermis, gets released due to disruption of the separating cell walls, resulting in the production of a homogeneous layer of the enzyme.

Rolling, the next step, macerates the leaf cells so that essential oils are reduced and the enzymes and their substrates get intricately mixed. One method to effect this is the use of orthodox roller, the rotorvane of the CTC system or their combinations, depending on the type of black tea needed i.e. leaf tea, granule tea or dust tea. The chemical changes that commenced during withering and preconditioning of the leaves, are accelerated at this stage due to temperature rise caused by frictional forces while rolling. Several *de novo* enzymatic reactions initiated during this process are continued right into the next phase, which is fermentation.

Fermentation, in reality, an oxidation process, continues for 45 minutes to 3 hours depending on the quality sought. Irrespective of its duration, the therapeutic products are generated mostly in this phase. Most important is the oxidation of the flavanols by polyphenol oxidase, which leads to the formation of a mixture of (–) epigallocatechin, (+) gallocatechin, (–) epictechin, (+) catechin, (–) epigallocatechin gallate and (–) epicatechin gallate (Roberts, 1962). These polyphenolic constituents of tea accountable for its quality, also act as scavengers of the damaging reactive oxygen species (ROS) responsible for the etiology of many chronic human health problems. The ultimate in this reaction series is the production of theaflavins and thearubigins and their fractions with characteristic molecular configurations. These compounds are responsible for both quality and pharmacological attributes of tea. The steps leading to the formation of theaflavins and thearubigins are illustrated in Fig. 1.1.

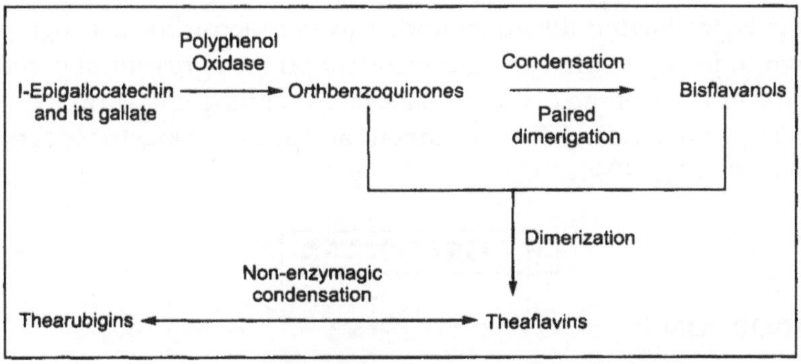

Fig. 1.1 Reactions during the fermentation process of tea

The last stage in the manufacturing process is firing, when the rolled and fermented leaf with a moisture content of 40-50% is dried to produce the black tea having a moisture content of 3%, by blowing hot air for about 20 minutes. The key aspect is the conversion of chlorophyll into pheophytin, which gives the processed tea the desired black appearance (Wickremasinghe and Perera, 1966). Moreover, the arrest of all enzymatic reactions during this stage ensures the stability of the therapeutic compounds generated earlier.

Finally, on the basis of particle size and type (whole or broken), the "fired" tea is physically sorted into different grades, but they continue to retain the chemical constituents ingrained earlier during processing. Common grades are Broken Orange Pekoe (BOP), Broken Pekoe (BP), Broken Orange Pekoe Fannings (BOPF), Flowrery Broken Orange Pekoe (FBOP), Fannings and Dust.

DISCUSSION

Despite the geographic variation of tea producing countries and variation in field practices in the production of tea, the manufacturing process remains fairly uniform and comparable. Consequently, the enzymatic reactions leading to the formation of the biologically active compounds during processing may not show any significant departure from the model presented here. The common feature in all geographic areas is that during processing the monomeric polyphenols that characterize the green leaf are polymerized into polymeric polyphenols with a reduction at approx. 60% in the level of biologically active compounds (Harbowy and Balentine, 1997). This reduction notwithstanding the therapeutic potentials of black tea remain unaffected (Krishnan and Maru, 2004). Basically the healthy difference is in the leaves.

However, the flavourous compounds may vary according to geography: the magnitude of variation being dependent on the agricultural practices and characteristics of the planting materials. Blending of a variety of teas from different areas also effect the flavour and aroma characteristics of tea (Owuor et al., 1988; 1990).

B. GREEN TEA

INTRODUCTION

Unlike black tea the world production of green tea is considerably low, the major producers being China, Taiwan, Japan and Indonesia. In India, green tea is produced in the Dooars, Darjeeling and parts of Tamil Nadu. Other production centres are in Himachal Pradesh, Uttaranchal and Tripura. Although world production of green tea has remained fairly steady (Table 1.3), production in India declined by 13.3% during 1997-1999. Yet a vast majority of research on human health benefits of tea was earlier based on green tea, possibly because catechins (beneficial nutrients) remained

Table 1.3 World production of green tea

(Quantity in '000 kgs)

Country	*1989*	*1990*	*1991*	*1999*	*2000*	*2001*
India	9879	8163	9822	6422	6216	5436
Bangladesh	948	1398	1278	242	372	186
Sri Lanka	964	909	805	389	950	1209
Indonesia	32500	34000	32950	37000	38000	39000
China	314,300	332,500	357,400	496,986	498,057	513,154
Japan	90500	89900	87900	88500	89300	89800
Vietnam	22300	23800	24400	32000	29000	32000
USSR	25400	26200	23200	–	–	–
Georgia	–	–	–	2200	2600	2700
Russian Federation	–	–	–	380	400	500
Thailand	3000	3000	3000	4000	4000	4000
China	89300	97500	100,800	129,986	137,973	145,596
Grand total	*589,091*	*617,370*	*641,555*	*798,105*	*806,868*	*833,581*

Source: Tea Board

virtually unaltered in the steaming process of green tea and hence provided opportunities for their pharmacological evaluation *in situ* (Hara, 2003; Takeo, 1992).

PROCESSING

The small size Chinary varieties or jats of tea leaves are generally used as raw materials in the production of green tea, though the use of large leaves is common in Indonesia. The main difference in the processing of green and black tea is the prevention of the fermentation (oxidation) process in the former by heat activation of the enzymes. This inactivation is achieved by either steaming the raw material or by "panning" or panfiring i.e. the material is fed into a hot rotating drum.

Steaming, the first step in green tea processing, ensures polyphenol oxidase and other enzymes are inactivated so that the green colour of the flush is maintained. Steam is injected for about 15-20 seconds into a cylindrical drum containing the leaves, followed by cooling of the treated leaves to make them ready for the next steps which are rolling and firing. Prior to that leaves are subjected to a treatment of hot air in a special chamber fitted with a rotating shaft to reduce the moisture content to about 50%.

Rolling under light pressure is done for 10 minutes followed by drying in a rotating drum for 20 minutes so that the moisture level of the leaf drops to

30%. A second and final rolling for 35 minutes gives the characteristic twist to the leaf, and a final drying with hot air stabilizes the moisture content at around 4%. A variant of the steaming process during manufacturing is "panning" though there are also other known forms of manufacturing depending on the characteristics needed for developing the local brands (Takeo, 1992).

BIOCHEMICAL CHARACTERISTICS

Green tea has its own specific flavour and aroma (Takeo, 1992). In general, the amounts of amino acids are higher in green tea than in black tea. Dimethylsulfide is an important constituent of the flavour of green tea, but other aroma inducing compounds have also been identified (Takeo, 1992).

From the perspective of therapeutics, the important difference between black and green tea is that in the former the polyphenols originally made up of four catechins, are oxidized into oligomers or polymers, and less than 5% of the catechin residues remain unoxidized, whereas in green tea, the four catechins forming the polyphenol matrix remain unoxidized in the absence of any fermentation (oxidation) during processing. The composition of the green tea catechins is presented in Table 3.4.

Table 3.4 Distribution of green tea catechins

Catechins	*Absolute percentage*
(+) gallocatechin	1.4
(–) Epigallocatechin	17.57
(–) Epicatechin	5.81
(–) Epigallocatechin gallate	53.90
(–) Epicatechin gallate	12.52

About 2 to 5% of the catechin is absorbed from the digestive tract of human beings and reach the plasma around 1 to 2 hours after ingestion (Miyazawa, 2001).

BIOLOGICAL PROPERTIES OF GREEN TEA

The potentially beneficial effects of green tea catechins are due to their antioxidant properties that sequester scavenging oxygen species and free radicals (Weisburger, 1999). EGCG in green tea extract, along with other catechins, can prevent tumerogenesis by scavenging the free radicals. EGCG is also anti-mutagenic, anti-angiogenic but pro-apoptotic (Sachindis and Hescheler, 2000).

Green tea polyphenols being capable of lowering the oxidation of low-density lipoprotein cholesterol, can decrease the risks of heart diseases (Weisburger, 1999). They can also effect a low ratio of triglycerides to high density lipoprotein, which is a good indicator of cardiovascular health, apart from controlling hypertension (Deng and Tao, 1998).

It is fairly well established that polyphenols of green tea can prevent several stages of carcinogenesis, including DNA damage and stabilization, cell proliferation, neoplastic growth and metastasis (Suganuma et al., 2001). More extensive human case studies are, however, needed to prove its unequivocal effects against cancer of the human lung, prostate, stomach and other vital organs. An interesting aspect, however, is that these polyphenols can stimulate the repairing of nucleotide excision by removing the DNA adducts produced by carcinogenesis. Consequently, green tea can and does enhance the actions of antitumour drugs (Tapier et al., 2002; Suganuma et al., 2001).

The antidiabetic effect of green tea polyphenols is basically due to their innate abilities to break down the starch and minimize the sudden rise in serum glucose. Other areas where green tea polyphenols could have preventive or curative effects, include bone erosions caused by infiltration of mono-nuclear cells and fibroblasts, and against arthritis by inhibiting the release of lysosomal enzymes.

The moot point is what quantity of polyphenol should be consumed to have a significant beneficial effect? It is generally believed that 3-4 gms of polyphenol, equivalent to 10 cups of tea per day, could be helpful. This, however, needs to be judged against the fact that half-life of epigallocatechin gallate, epicatechin gallate and epicatechin is 191, 362 and 45 minutes respectively, and they are the principal constituents of green tea polyphenols (Zhu, Chen and Li, 2000).

REFERENCES

Banerjee, B.1992. Botanical classification of tea. In: Tea Cultivation to Consumption (eds. K.C. Willson and M.N. Clifford). Chapman and Hall, London, U.K. pp 25-51.

Banerjee, B. 2001a. Tea. In: Ecosystem of the World (Vol. 19): Tree Crop Ecosystem (ed. F.T. Last) Elsevier, Amsterdam, The Netherlands. pp 361-378.

Banerjee, B. 2001b. Can plucking increase harvest index of tea? J. Plant. Crop. 22: 66-67.

Banerjee, B. 2002. Tea Production and Processing (4th ed). Oxford & IBH Publishing Co. Pvt. Ltd., New Delhi, India.

Barua, D.N. 1987. Tea. In: Tree Crop Physiology (eds: M.R. Sethuraj and A.S. Raghavendra). Elsevier, Amsterdam, The Netherlands. pp 225-226.

Deng, Z.Y. and B.Y. Tao. 1998. Effect of green tea and black tea on blood glucose, triglycerides and antioxidants in aged rats. J. Agric. Food. Chem. 46: 3875-3878.

Graham, H. 1984. Tea: The Plant and Its Manufacture, Chemistry and Consumption of the Beverage. In: The Methylxanthine Beverages and Food (ed: G. Spiller). L.R. Allian Inc. New York, USA. pp 29-44.

Hara, Y. 2003. Tea as a preventive/protective agent against human diseases. Int. J. Tea Science 2: 20-26.

Harbowy, M. and D. Balentine. 1997. Tea chemistry. Crit. Rev. Plant Sciences 16: 415-480.

Krishnan, R. and G.B. Maru. 2004. Inhibitory effect(s) of black tea polyphenol fractions on the formation of (3H) – B(a) P-derived DNA adducts. J. Agric. Food. Chem. 52: 4261-4269.

Miyazawa, T. 2001. Absorption and metabolism of tea catechins in humans: Proc. 2001 Int. Conf. on O-cha (tea) Culture and Science. III: 34-37.

Owuor, P.O., T. Tsushida, H. Horta and T. Murai. 1988. Effect of geographic area of production on the composition of volatile flavour compounds in Kenyan clonal black teas. Expt. Agric. 24: 227-235.

Owuor, P.O., S.O. Obega and C.O. Othieno. 1990. Effects of attitude on the chemical composition of black tea. J. Sci. Food. Tech. 50: 9-19.

Roberts, E.A.H. 1962. Economic importance of tea flavonoid substances: tea fermentation. In: The Chemistry of Flavonoid Compounds (ed. T.P. Geissman). Macmillan & Co., New York, USA. pp 468-512.

Sachindis, A. and J. Hescheler. 2002. Are catechins natural tyrosine kinase inhibitors? Drug News Perspect. 15: 432-438.

Suganuma, M., Y. Ohkura, S. Okabe and H. Fujiki. 2001. Combination cancer chemoprevention with green tea extract and sulindac shows in intestinal tumor formation in min mice. J. Cancer Res. Clin. Oncol. 127: 69-72.

Takeo, T. 1992. Green and semi-fermented tea. In: Tea Cultivation to Consumption (eds: K.C. Willson and M.N. Clifford). Chapman and Hall, London, UK. pp. 413-457.

Takeo, T. and P.K. Mahanta. 1972. Comparison of black tea aroma of orthodox and CTC teas, and black teas of different varieties. J. Sci. Food Agric. 23: 227-232.

Tapiero., H., K.D. Tew, N. Ba and G. Matha. 2002. Polyphenols: Do they play a role in the prevention of human pathologies? Biomed. Pharmacother. 56: 2000-2007.

Weisburger, J.J. 1999. Mechanism of action of antioxidants as exemplified in vegetables, tomatoes and tea. Food. Chem. Toxicol. 37: 943-948.

Trevisanto, S. and Y.I. Kim. 2000. Tea and health. Nutr. Rev. 58: 1-10.

Wickremasinghe, R.L. and K.P.W.C. Perera. 1996. The blackness of tea and the colour of tip. Tea Q.37: 75-79.

Wood, D.J. and P.K. Barua. 1958. Species hybrids of tea. Nature, 181: 1674-1675.

Wood, D.J. and N.B. Chanda. 1955. Annual Report, Biochemistry Branch, Tocklai Experimental Station. 43-64.

Zhu, M., Y. Chen and R.C. Li. 2000. Oral absorption and bioavailability of tea catechins. Plants Medica 66: 444-447.

Chapter

2

The Diversity of Therapeutic Compounds in Tea

B. Banerjee
National Tea Research Foundation, Kolkata, India

Tea has interesting chemical constituents, some of which are as unique as the plant itself. Like the chequered history of tea, the chemical history of tea is quite fascinating, to say the least. From the dawn of commercial planting of tea in India, attention has been paid to identify what is the basic element that is attributable to the quality of tea, though earlier the search was based on the assumption that a single influencing factor or ingredient could be responsible for the quality of tea (Ukers, 1935)

Slowly but surely, the perspective changed with advances in chemical research and the main constituents of tea leaves were broadly identified to be carbohydrates, proteins, polyphenols, caffeine, theanine, vitamins and minerals (Wickremansinghe, 1979). But these chemicals in tea leaves (Appendix 1) undergo dynamic changes as green tea leaves are converted into black tea during the process of manufacturing. There is, therefore, nothing static about these compounds because they undergo rapid changes not only during manufacturing process, but also according to agricultural practices, shade from sunlight or its absence, plant location, season, and of course, leaf types or varieties (Banerjee, 1992b). The leaf also contains enzymes that facilitate the chemical reactions in processing young shoots to produce green, black and oolong tea (Banerjee, 2002).

From therapeutic point of view, polyphenols and caffeine are pivotal, though vitamins and minerals are also of considerable medicinal significance (Chakravarty, 1990). It is generally believed that these chemical constituents of tea, together or separately, and most probably by their interactions, affect both its quality and medicinal attributes (Stagg and Millin, 1975).

Polyphenols

Tea contains 30 to 42% polyphenols on a dry-weight basis, catechins being particularly important. Catechins generally belong to a group of compounds, the flavonoids, which have C_6-C_3-C_6 carbon structure with two aromatic rings (Fig. 2.1). Catechins in tea are unique, as a kind of them are not akin in any other plants (Haslam, 1989).

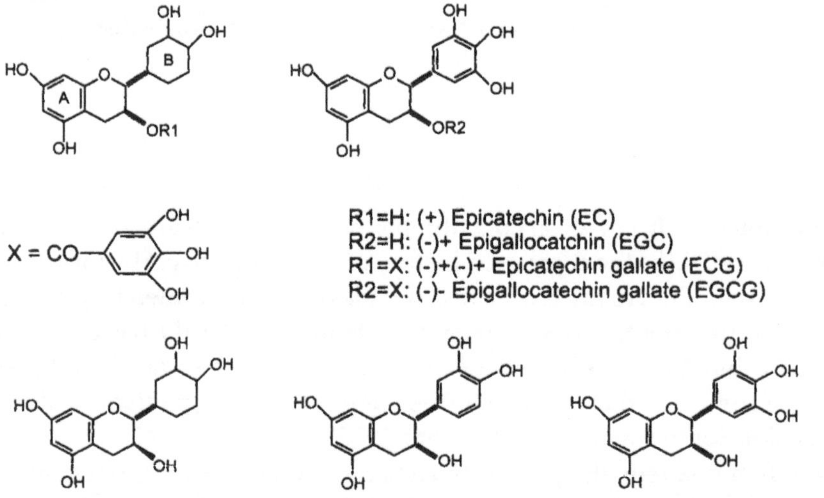

(+)-Catechin (+C) (+)-Epicatechin (EC) (-)+Epigallocatchin (EGC)

Fig. 2.1 Structures of some compounds of pharmaceutical importance.

Catechin derivatives of primary polyphenols were first reported using the technique of paper chromatography by Roberts and Wood (1950). These include

(i) (–)-epigallocatechin-3-gallate (EGCG)
(ii) (–)-epigallocatechin (EGC)
(iii) (–)-epicatechin-3-gallate (ECG)
(iv) (–)-epicatechin (EC)

The highest concentration is that of EGCG, followed by EGC, ECG and EC in decreasing order. Other catechins include:

(i) (+)-gallocatechin (GC)

(ii) (–)-gallocatechin gallate (GCG)
(iii) (–)-catechin gallate (CG)
(iv) (+)-catechin

These are present in small qualities.

A cup of tea may contain between 300 and 400 mg of polyphenols in totality. EGCG is the major polyphenolic constituent, contributing about 25 to 40% of the total catechin load of tea.

During the biosynthesis of catechins, C_6(A) catechin ring is produced by the acetc-malonic pathway and C_3-C_6 (B) by the shikimic-cinnamic acid pathway starting from the glucose pool (Zaprometov, 1963). Synthesized catechins are stored in cell vacuoles without undergoing any further metabolism. (–) Epigallocatechin is produced by hydroxylation of (–) epicatechin. (–) Epicatechin gallate and epigallocatechin gallate are, however, synthesized by the esterification of catechins with gallic acid present in the cell (Iwasa, 1977).

Polyphenols possess high antioxidant properties and protect human cells from adverse effect of damaging reactive oxygen species (ROS). The etiology of most human diseases can be traced to ROS, and in the event polyphenols act as scavengers to prevent damages to cellular macromolecules (Dufresne and Farnworth, 2001).

The catechin content of tea varies according to the genotype. The Assam variety generally accumulates more catechins than the China variety. In the former, catechin accounts for 30% of the dried matter against 20% in other varieties.

Roberts et al. (1957) revealed two-dimensional paper chromatograms with two orange coloured spots and one reddish brown tailing spot. The former with its gallate was named "theaflavin" and the latter "thearubigin". Individual catechins (precursors) mixed with polyphenol oxidase also resulted in the development of bisflavanols in addition to theaflavins. In theaflavins, Roberts et al. (1957) confirmed the existence of benzotropolone structure. Crystals were obtained by reacting (–) EC and (–) EGCG in the presence of polyphenol oxidase which were the same compound as theaflavin. But polyphenol oxidase takes part only in the initial formation of quinones and reactions show the presence of four main kinds of theaflavins that constitute 1 to 2% of black tea on a dry-weight basis.

Thearubigin is formed by the polymerization of catechins and constitute 10 to 20% of black tea which is ten to twenty times greater than the dry weight of the theaflavins. Thearubigin is a heterogenous group of compounds with a molecular weight of 700 – 40,000. The degradation of thearubigin fraction leads to the formation of flavan-3-ols, flavan-3-ol gallates, anthocyanidin, delphinidin and gallic acid, as well as polysaccharide and protein residues. Thus during fermentation, catechins, proanthocyanidins and degraded theaflavins might form a dialysable low molecule thearubigin at first, and then further polymerisation and complexation with protein or polysaccharide form undialysable thearubigin. A high theaflavin content usually implies good manufacturing practices.

Caffeine

Caffiene (trimethylxanthine), a purine alkaloid constitutes 2.5 to 5.5% of the total chemical constituents of tea leaf on a dry-weight basis and occurs along with small quantities of dimethylxanthine, monomethylxathine and theophylline (Wood et al., 1964). Caffeine, being a product of the methylation of purine nucleotides, does not result from the degradation of ribonucleic acid, as was supposed to be earlier (Suzuki and Takahashi, 1976). Recently, a gene in the tea leaf encoding caffeine synthase, N-methyltransferase, has been cloned and the recombinant enzyme produced in *E. coli* (Ashihara and Crozier, 2001) has opened the possibilities of generating caffeine-free tea plants.

The effect of caffeine on the human system is complex. It stimulates the brain and the nervous system signalling the production of the hormone adenosine which slows down the rate of metabolism, a kind of sedative in effect. It is of interest to note that caffeine and its metabolites do not accumulate in the body as they are generally demethylated, oxidized and excreted (Marks, 1992).

Vitamins

Tea contains a wide range of vitamins that include carotene, riboflavin, nicotinic acid, pantothemic acid, ascorbic acid and theamine, though most of them are lost during the processing of fresh green tea leaves to black tea. Theanine, which is γ-ethylamide of glutamic acid having an antagonistic action against the stimulating action of caffeine, is also unique to tea. Vitamin B content of black tea ranges from 1.4 μg of bioten to 127.5 μg of riboflavin.

Carbohydrates

Carbohydrates constitute about 4-5% of the solids extracted in tea infusion. Though concentrated mostly in the roots, carbohydrates are not uncommon in the leaves. From nutritional point of view, carbohydrate content is significant (Chakravarty, 1990).

Lipids

Lipids are an interesting group that show marked variation in their chemistry. Young tea leaves, for example, contain more phosphatidyl ethanolamine and phosphatidyl choline than mature leaves, but monogalactosyl diglyceride and digalactosyl glyceride are present in higher concentrations in mature leaves (Roberts, G.R., 1974; Takeo and Tsushida, 1980).

Triterpenoids

Of the two fractions in tea, the unsaponifiable portion has spinasterol glycoside and β-amyrin in leaves, root and stems. The other fractions, the saponins, are mostly active in the roots (Wickremasinghe, 1979).

Carotenoids and Pigments

The fourteen carotenoid compounds identified in tea leaves (Wickremasinghe, 1979) are mostly concentrated in mature leaves. Whether they have any therapeutic significance is not clear. Tea pigments, however, have considerable medicinal properties – particularly beneficial in cardiac diseases. 125-250 mg of green tea pigments thrice per day appear to have good effect on human subjects (Banerjee, 1992a).

Minerals

Minerals constitute about 4 to 9% of the inorganic matter of tea. Of particular interest for dental health is the fluorine (3-200 ppm) in leaves which could prevent dental decay. The role of aluminium (20-11000 ppm) is controversial, but potassium (9000-34000 ppm) could have a supplementary effect in diets lacking potassium (Banerjee, 1992a). Using the techniques of principal component analysis, linear discriminate analysis and artificial neuron network, these minerals have also been found useful in locating the geographic bases of tea varieties used in blending tea (Fernandes-Caceres et al., 2001).

Overall Distribution of Compounds

All parts of the tea plant of any variety store chemicals, though they vary qualitatively as well as in composition. Compounds found in different parts of tea (Table 2.1) would give an idea of the diversity of the compounds in different plant parts (Banerjee, 2002).

Table 2.1 Distribution of chemical compounds in different parts of tea

Compounds	*Young leaf*	*Mature leaf*	*Green stem*	*Mature stem*
Polyphenols	++	+	+	+
Amino acids	++	+	+	+
Nucleotides	+	+	+	+
Phosphate esters	+	+	ND	ND
Caffeine, theobromine	++	+	+	+
Carbohydrates	+	+	+	+
Lipids	+	+	+	ND
Organic acids	+	+	+	+
Chlorophyll	+	+	+	+
Carotenoids	+	+	+	+
Unsaponifiables	+	+	+	+
Saponin	+	+	+	+
Minerals	+	+	+	+
Volatile compounds	+	+	+	+

+ = relative abundance; ND = not determined

REFERENCES

Ashihara, H. and A. Crozier. 2001 Caffeine: A well-known but little mentioned compound in plant tissues. Trends Plant.Sci.6: 407-413.

Banerjee, B.1992a. Tea As A Health Drink. Coordination Committee of the Planters Association. Kolkata.

Banerjee, B. 1992b. Botanical classification of tea. In: Tea Cultivation to Consumption (eds. K.C. Willson and M.N. Clifford). Chapman and Hall, London, U.K. pp 25-51.

Banerjee, B. 2002. Tea Production and Processing. Oxford & IBH Publishing Co. Pvt. Ltd., New Delhi (3rd edition).

Chakravarty, I. 1990. Nutritional value of tea. In: Tea Research: Global Perspective (ed. B. Banerjee). Tea Research Association, Kolkata. pp. 163-167.

Dufrense, C.J. and E.R. Farnworth. 2001. A review of latest research findings on health promotion properties of tea. J. Nutr. Biochem. 12: 404-421.

Fernandes-Caceres, P.L., M.J. Martin, F. Pablos and A.G. Gonzalez. 2001. Differentiation of tea (*Camellia sinensis*) varieties and their geographic origin according to their metal content. J. Agric. Food Chem. 49: 4775-4779.

Haslam, E. 1989. Plant Polyphenols. Vegetable Tannins Revisited. Cambridge University Press, Cambridge.

Iwasa, K. 1977. Biosynthesis of catechins in tea plant. Bull. Natl. Res. Inst. of Tea 13: 101-126.

Marks, V. 1992. Physiological and clinical effects of tea. In: Tea Cultivation to Consumption (eds: K.C. Willson and M.N. Clifford). Chapman and Hall, London, U.K. pp. 707-739.

Roberts, E.A.H. and D.J. Wood. 1950. The fermentation process in tea manufacture. II. Oxidation of substrates by tea oxidase. Biochem. J. 47: 175-186.

Roberts, E.A.H., R.A. Cartwright and M. Oldschool. 1957. The phenolic substances of manufactured tea. I. Fractionation and paper chromatography of water-soluble substances. J. Sci. Food Agric. 8: 72-80.

Roberts, G.R. 1974. Polar lipid composition of the leaves and seeds from the tea plant (*Camellia sinensis* L.). J. Sci. Food Agric. 25: 473-475.

Stagg, G.V. and D.J. Millin. 1975. The nutritional and therapeutic value of tea. J. Sci. Food Agric. 26: 1439-1459.

Suzuki, T. and E. Takahashi. 1976. Caffeine Biosynthesis in *Camellia sinensis*. Phytochemistry, pp. 1235-1240.

Takeo, T. and T. Tsushidi. 1980. Changes in lipoxygenase activity in relation to degradation in plucked tea shoots. Phytochemistry, 19: 2521-2522.

Ukers, W.H. 1935. All About Tea. The Tea and Coffee Journal Limited, New York. USA.

Wickremasinghe. R. 1979. Tea. In: Advances in Food Research. Vol. 24 (eds. E. Mark and G.F. Steward). Academic Press, New York. pp. 229-286.

Wood, J., I.S. Bhatia, S. Chakravarty, M.D. .Chaudhary, E.A.H. Roberts,and M. Ullah, 1964. The chemical basis of quality in tea. Analysis of freshly plucked shoots. J. Sci. Food Agric. 15: 8-14.

Zaprometov, M.N. 1963. Catechin biosynthesis in tea shoots. Fiziol. Rast. 10: 73-78.

Chapter

3

Production of Therapeutic Compounds in Tea

T.C. Chaudhuri
Tea Board, Kolkata, India

The antioxidant properties of tea are fairly well established, though more than tea leaf per se, it is the chemical compounds within the leaf that are of consequence. Most *in vitro* studies have used tea extracts in evaluating the effects of tea against simulated human diseases. It would therefore be worth noting the compounds that are present in fresh green leaf compared to the processed leaf, and the chemical transformations as the green leaf is processed into black tea.

POLYPHENOLS

From therapeutic point of view, the polyphenols in tea leaves are most important, despite the presence of many other compounds in the leaf (Banerjee, 2005). Young tea shoots are rich in polyphenolic compounds, the largest component being those of catechins constituting about 30% of the total chemical biomass on dry-weight basis (Roberts, 1962).

Flavanol, the other important group, is present at a level of 1% or more in the leaf, again on dry-weight basis. The flavanols identified in the green leaf are:

1. (+) catechin
2. (+) gallocatechin
3. (–) epicatechin

4. (–) epigallocatechin
5. (–) epicatechin-3-gallate
6. (–) epigallocatechin-3-gallate.

Two other compounds reported from fresh green the leaf (Saijo, 1982) are:

7. (–) epicatechin -3, 5-digallatene
8. 3-methyl digallate of (–) epicatechin and (–) epigallocatechin.

In processed tea, i.e., the tea that has turned into black from green after going through withering, rolling, fermentation and drying, the first six compounds continue to be present in varying proportions. Additionally two others compounds present are

9. (+)-catechin-3-gallate
10. (+) gallocatechin-3-gallate

The last two compounds, according to Roberts (1962), are the products of epimerization during drying, hence of a post-fermental origin.

Variations in Specific Compounds

One can possibly look at freshly harvested tea shoots for the distribution of polyphenols and then note the quantitative changes in the individual compounds during processing. This is important because for therapeutic purposes, the black tea extract (BTE) containing the penultimate compounds is mostly the working substrate. The distribution of polyphenols in fresh shots decline with leaf age as shown in Table 3.1.

Table 3.1 Distribution of polyphenol in the tea shoot

Part	*% Dry weight*
Bud	35.8
1 leaf	35.0
2 leaves	27.9
3 leaves	23.1
Petiole	15.0

Clearly, polyphenol content declines with the age of the leaf. But, as Wickremasinghe and Perera (1973) point out the proportion of individual catechins relative to each other would also vary in an age-dependent manner.

Changes During Processing

During withering the water content of the harvested shoots is reduced to around 16% but a process of chemical withering starts almost from the time of plucking. It is not clear, whether this is the point for generation of catechins, or the products of therapeutic importance. However, in the next step, i.e. rolling, disruption of cells ensures release of enzymes, particularly polyphenol oxidase, that initiates a series of chemical reactions culminating in the serial production of catechins, and other compounds of importance in tea flavour and quality.

Epigallocatechin and epigallocatechin gallate are generally present in fresh green leaves in higher concentrations than epicatechin and epicatechin gallate (Bhatia and Ullah, 1968). The ratio, however, starts changing with withering, and rolling sets the trend towards catechin concentration in the final product (Bhatia and Ullah, 1968).

The process of serial chemical reaction continues into the fermentation process when the enzymatic oxidation of the green leaf catechins ultimately leads to the formation of theaflavins and thearubigins (Robertson and Bendal, 1983).

THEAFLAVINS AND THEARUBIGINS

These two major chemical complexes of tea are of great significance both for quality and medicinal attributes. They are basically products of oxidation of catechins (Roberts, 1950), which were identified by chromatographic studies. Additionally, a gallate of theaflavin, namely, theaflavingallate, was also identified (Roberts, 1950).

Theaflavins are chemically complex with the identifiable fractions (Coxon et al., 1970). The identifiable fractions and their proportions in black tea are shown in Table 3.2.

Table 3.2 Fractions of theaflavins and their distribution

Fraction	*% distribution*
Theaflavin	18
Theaflavin-3-gallate	18
Theaflavin-3′-gallate	20
Theaflavin-3-3-digallate	40
Isotheaflavin + theaflavic acid	4

The proportion of the fractions of thearubigins, however, varies according to fermentation time and possible plant genotypes (Robertson, 1983).

Thearubigins, responsible for the colour of tea infusion, have complex configurations, and account for 10-20% of the chemical constituents of black tea, on dry-weight basis (Roberts, 1950). Five fractions of thearubigins have been isolated from tea liquors (Brown et al., 1969).

Each fraction of thearubigin contains 18-26% of anthocyanidins and these fractions, in general, have a structure base of polymeric proanthocyanidins (Brown et al., 1969). This view has, however, been disputed (Opie et al., 1990), though there is little doubt that thearubigins are directly derived from the catechins as well as from the breakdown of theaflavins and theaflavin intermediaries (Robertson, 1983). Overall, the dynamics of thearubigin is still a contested subject.

Discussion

The compounds of medicinal importance are no doubt universally present in tea extracts, though proportions of each compound in the extract are not well documented. When a particular tea infusion is proved effective, it is often difficult to apportion the effect to any specific compound or a range of compounds in tea extracts.

There are studies involving evaluation of individual catechins (Luthria et al., 1997). But comparative studies with a range of tea compounds against a specific disease can only indicate the relative or even absolute disease-specific therapeutic potentials of these compounds.

A clarity still lacks as to the methodology to be adopted in therapeutic evaluation of tea. Polyphenols have been isolated and commercially produced for the human health of benefits. Overall, however, the emphasis is more on BTE containing approximately 20% polyphenol, more precisely, the compounds derived from them during processing. It is, therefore, the synergism of these compounds in BTE that appears crucial in acting against disease causing organisms both in dosage and time dependent manner. Unquestionably, polyphenols are vital in human health, but when claims are made for their universal effectiveness, it would be helpful to know the causal pathway towards amelioration or prevention of diseases (Yoshino et al., 1994).

REFERENCES

Banerjee, B. 2005. The diversity of therapeutic compounds in tea. pp. 15-21. (this volume).

Bhatia, I.S. and M.R. Ullah. 1968. Polyphenols of tea. Studies with different organs and some cultivated varieties of tea. J. Sci. Food Agric. 19: 535-542.

Brown, A.G., W.B. Yton, A. Holmes and W.D. Olis. 1969. Identification of thearubigins as polymeric proanthocyanidins. Nature, 221: 742-744.

Coxon, D.T., A. Holmes, W.D. Ollis and V.C. Vora. 1970. Constitution and configuration of the theaflavin pigments of black tea. Tetrahedron Lett. pp. 5237-5240.

Luthria, D.L., A.D. Jones, J.L. Donovan. and A.L. Waterhouse. 1997. Determination of catechin and epicatechin levels in human plasma. J. High Resol. Chromatography. 20: 621.

Roberts, E.A.H. 1950. The phenolic substances of manufactured tea II. Thearubigins as enzymatic oxidation products in fermentation J. Sci. Food and Agric. 9: 212-216.

Roberts, E.A.H. 1962. Economic importance of flavanoid substances. Tea fermentation. In: The Chemistry of Flavonoid Substances (ed. T.A. Geissman). Pergamon, Oxford. pp. 468-512.

Saijo, R. 1982. Isolation and chemical structures of two new catechins from fresh tea leaf. Agric and Biol. Chem. 46: 1969-1970.

Robertson, A. 1983. Effect of catechin concentration on the formation of black tea polyphenols during *in vitro* oxidation. Phytochemistry, 22: 897-903.

Robertson, A. and D. Bendal. 1983. Production and HPLC analysis of black tea theaflavins and thearubigins. Phytochemistry, 22: 883-887.

Opie, S.C., A. Robertson and M.N. Cliffor. 1990. Black tea thearubigins – their HPLC separation and preparation during *in vitro* oxidation. J. Sci. Food Agric. 50: 547-561.

Wickremasinghe, R.L. and K.P.W.C. Perere. 1973. Factors affecting quality, strength and colour of black tea liquors. J. Nat. Sci. Coun. Sri Lanka. 1: 111-121.

Yoshino, K., Y. Hara, M. Sano and I. Tomita. 1994. Antioxidative effects of black tea theaflavins and thearubigin on lipid peroxidation of rat liver homogenates induced by tent-butyl hydroperoxide. Biol. and Pharmacol. Bull., 17: 146-149.

Chapter

4

Pharmacological Evaluation of the Medicinal Properties of Tea

D.K. Roy, A.K. Nag Chaudhuri and T. Sen
School of Natural Product Studies
Jadavpur University, Kolkata, India

Tea (*Camellia sinensis*) is one of the most commonly consumed beverages in the world today. Since time immemorial, tea has been regarded as a healthy beverage. In ancient Chinese literature (from Tang Dynasty), tea has been rated as a leading health giving beverage and cure for many diseases. According to a report published by Jelinck (1978), the consumption of tea infusion was probably popular even 500,000 years ago. As per the available information, LuYu authored the first book on tea during 780 A.D.

The different varieties of tea (Green, Black and Oolong) were possibly introduced during the 14th century (Ming Dynasty). The Darjeeling variety of tea was first cultivated during 1841; however, commercial production experienced a major growth during the second half of the 19th century.

The health benefits of tea have been known to human civilization for centuries. However, when we take a look into the historical aspect related to tea, we find that in earlier days tea infusion was popular for improving blood flow, detoxification and disease prevention (Balentine et al., 1977). However, on the basis of scientific literature, only a few reports were compiled on the medicinal values of tea up till 1990 (Fig. 4.1). As is also evident from Fig. 4.1, the research activity on tea gathered tremendous momentum only during the last decade.

The types and percentage content of flavonoids (polyphenols) present in tea, usually differ depending on the variety of leaf, environment in which tea is grown, processing, manufacturing, particle size of ground tea leaves and

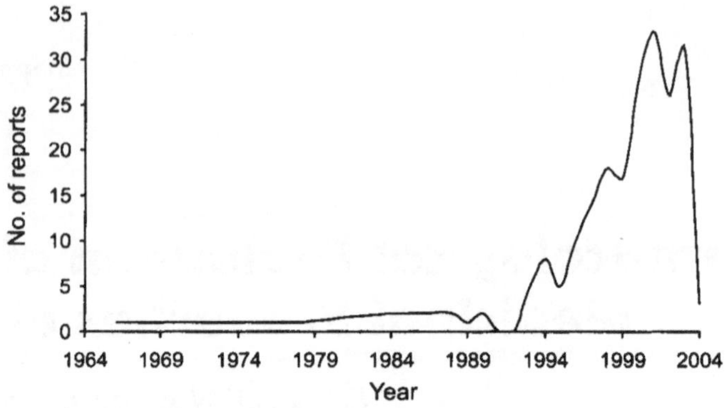

Fig. 4.1 An overview of number of scientific reports on black tea (1966-2004)

infusion preparation. Typically, more than 90% of total tea phenolic compounds are reported to be flavonoids.

Based on the manufacturing procedures, tea can be broadly classified into three major categories:

1. Green tea is prepared by steaming freshly plucked tea leaves in large vats. The process of steaming is useful for preserving the green colour. Green tea contains several polyphenols, which include flavanols, flavandiols, flavonoids and phenolic acids; these compounds may account for up to 30% of the dry weight. Most of the green tea polyphenols are flavonols, commonly known as catechins. Some major green tea catechins are:
   (–)-epigallocatechin-3-gallate (EGCG),
   (–)-epigallocatechin (EGC),
   (–)-epicatechin-3-gallate (ECG),
   (–)-epicatechin (EC),
   (+)-gallocatechin and
   (+)-catechin.
   Green tea also contains Caffeine, theobromine and theophylline, the principal alkaloids, account for about 4% of the dry weight. In addition, there are phenolic acids such as gallic acids and amino acids such as theanine.
2. Oolong tea is derived from partial fermentation of fresh leaves (greenish brown colour). Oolong tea, a partially oxidized tea, contains monomeric catechins, theaflavins and thearubigins. Some other characteristic components found in Oolong tea are: epigallocatechin esters, theasinensins, dimeric catechins and dimeric proanthocyanidins.

3. Black tea – the fresh mature leaves are subjected to a complete oxidation step prior to drying. These leaves are procured from the tea garden of Assam, Darjeeling and also from Sri Lanka and Kenya. The process of preparation includes the following stages:
   (i) crushed
   (ii) torn
   (iii) curled

In Black tea, the monomeric flavan-3-ols, undergo polyphenol oxidase-dependent oxidative polymerization leading to the formation of bisflavanols, theaflavins, thearubigins, and other oligomers. Theaflavins (about 1%-2%) of the total dry matter of black tea, including theaflavin, theaflavin-3-O-gallate, theaflavin-3′-O-gallate, and theaflavin-3,3′-O-digallate (possessing the benzotropolone rings) is known to impart the characteristic colour, and taste of black tea. About 10-20% of the dry weight of black tea is contributed by thearubigens, which have a wide range of molecular weights (Yang et al., 1993).

Although, not having a replacement for fruits and fresh vegetables, the antioxidant activity of tea has been compared to that of fruits and vegetables in numerous studies. One report (in U.K.) concluded that daily consumption of 3 cups of tea has approximately the same antioxidant power as that of six apples (Paganga et al., 1999). Another study confirmed that one or two cups of tea have the same 'radical scavenging capacity' as five portions of fruit and vegetables or 400 mg Vitamin C equivalents (Du Toit, 2001). Drinking tea in a diet with a combination of vegetables and fruits can be revitalizing to health.

Researchers have proved that tea drinking can decrease the risk of heart diseases, cancers of lung, pancreas, breast, stomach, liver, colon and prostrate, hypertension and brain stroke (Trevisanto et al., 2000). They have proved that the antioxidant properties of tea can inactivate the influenza virus, and also decrease cholesterol and low-density lipoprotein (Osada et al., 2001). Despite of all the reports, the Indian tea industry is yet to capitalize on these new findings. Thus information related to health benefits of tea, when conveyed to the public in general and the medical practitioners may bring about massive changes in the consumption pattern.

As is evident from extensive literature search, the pattern of research activity with respect to tea, is primarily centred on the Cardiovascular System (CVS), cancer therapy and the gastro intestinal tract (Fig. 4.2), however, studies are also being carried out for evaluating other pharmacological properties of tea.

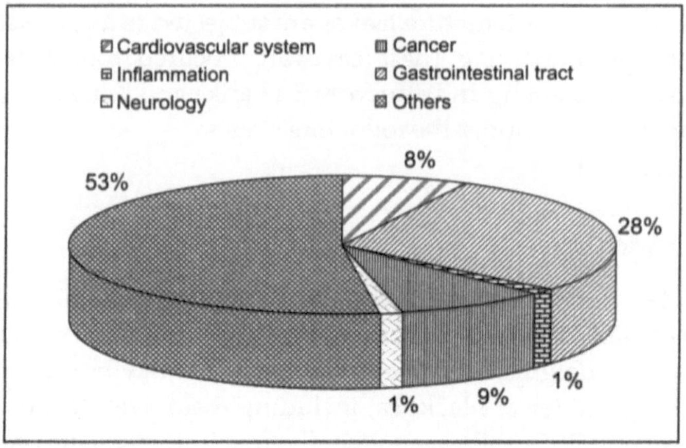

Fig. 4.2 Pharmacological avenues explored for activities of black tea extract since1966

PHARMACOLOGICAL ASPECTS ASSOCIATED WITH TEA CONSUMPTION

Cardiovascular System

A number of epidemiological studies (Sesso et al., 1999; Keli, 1996; Knekt et al., 1996; Stensvold, 1992; Yochum et al., 1999; Hertog et al., 1993; Hertog et al., 1995; Geleijnse et al., 1999) indicate that the total polyphenol content following tea consumption can be linked to protection of the Cardiovascular system (CVS). However, some other studies have reported dissimilar results with respect to protective effects as discussed earlier. Such inconsistencies from these studies could be arising due to a number of possibilities such as (also see p. 187 in this volume):

(a) quantity of tea consumed per day
(b) strength of the brew
(c) variety
(d) source
(e) flavonoid contents
(f) genetic variations

Pharmacological and clinical evidences indicate that the tea polyphenols could be beneficial in lowering the risk of cardiovascular disease (CVD) associated with high cholesterol levels and hypertension. Increased plasma antioxidant level (persons drinking 3-4 cups of tea per day) is known to reduce the possibility of myocardial infarction (Hertog et al., 1997).

From different studies (in vitro and in vivo), carried out with tea, it is evident that flavonoids and other polyphenols present in tea may protect the cardiovascular system by prevention of LDL oxidation (Rimm et al., 1996; Woodward, 1999; Klatsky et al., 1990; Peters et al., 2001; Vinson, 1995; De Whalley et al., 1990), lowering serum lipids (Yoshida et al., 1999; Pearson, 1998; Zhu et al., 2000), anti-inflammatory actions (Osada et al., 2001; Yang and Kool, 1997), improvements in blood vessel function (Lin, 1998) and inhibition of platelet aggregation (Vinson et al., 1998; Tijburg et al., 1997; Hofbauer et al., 1999; Duffy, 2001; Corvazier et al., 1985). Clinical studies on human subjects are not conclusive, although some epidemiological studies seem to indicate that tea can reduce serum cholesterol (Polette et al., 1996). Interestingly, some studies associated with Cardiovascular Disease (CVD), however, show insignificant activity of tea in reducing serum cholesterol and blood pressure (Dufrense and Farnsworth, 2001).

A study has indicated that both short and long term consumption of black tea reverses vasomotor dysfunction in patients with congestive heart disease, thus indicating that tea may impart significant protection to the cardiovascular system (Duffy et al., 2001).

Tea (particularly catechin and theanin) is known to play an important role in renal hypertension and progression of renal failure (Yokozawa et al., 1994,1996 and 1998).

CANCER

There is increasing evidence that specific compounds present in certain natural foods, are vital requirements for human nutrition. These foods, if included in the regular diet, can improve the healthy lifestyle. Tea is rich in specific phenolic compounds including flavonoids, which are known to be powerful antioxidants and have been reported to possess anticancer properties (Middleton, 1994; Smith, 2001; Lynn-Cook et al., 1999). In a population survey (conducted in Japan), the study indicated towards an overall protection coupled with a decrease of cancer incidences, in individuals habituated to tea drinking. The effects were more prominent when the consumption rate was increased (per day).

In some reports it was observed that there was a decrease in the incidence of stomach cancer with increased consumption of tea (Setiawan et al., 2001; Yu et al., 1991; Yu et al., 1995; Ji et al., 1996; Inoue et al., 1998; Kono, 1988; Bushman, 1998), and according to one report, it was observed that green tea drinkers had experienced a 48% reduction in the risk of developing stomach cancer. Studies indicate that tea consumption (polyphenols) can inhibit tumorigenesis at its initiation, and also during progressive stages of cancer (Dreosti, 1997; Katiyar, 1996). It has now been considered that such anti-cancer activity can be attributed to:

(a) Antioxidant activity (Xu et al., 1992; Wang et al., 1989)
(b) Inhibition of nitrosamine reactions (Wang et al., 1992)
(c) Modulation of carcinogen-metabolising enzymes (Jakun, 1997)
(d) Ability to inhibit cell-proliferation (Yang et al., 1993)
(e) Modulation of gut-microflora (Yamamato, 1997) associated to colon cancer
(f) Possible antimicrobial (*Helicobacter pylori* in the gastric tissue) actions (Wisburger, 1999; Diker et al., 1994)

A study conducted in Japanese population indicates decrease of cancer incidences, associated with tea drinking (Imai et al., 1997). According to some other studies, green tea drinkers had a 46% reduction in the risk of developing stomach cancer and a 51% lowered risk of developing chronic gastritis as compared to non-drinkers (Setiawan et al., 2001). Flavonoids have also been associated with a reduced risk of lung cancer and in one case control study, it was observed that there was a 50% reduction in the risk of lung cancer following consumption of one cup of black tea daily (Knekt et al., 1997). However, findings from investigations have not been able to establish any relationship between catechin intake (a type of flavonoid that is abundant in tea) and lung cancer (Arts et al., 2001). A recent study conducted in Japan showed that regular consumption of green tea (more than 3 cups a day) could be effective during the initial stages of breast cancer (Inoue et al., 2001). Although, researchers claimed that such results need to be carefully analysed and further studies are necessary to confirm these findings.

With respect to different stages of cancer development, it can be mentioned that tea polyphenols particularly EGCG and ECG are effective during the initiation process, EGCG and thealavin during the promotion stage and finally EGCG, EGC, ECG, theflavins and other tea polyphenols are effective during the progression stage (Pillai et al., 1990; Chen et al., 1999; Fujiki et al., 1999; Naasani et al., 1998; Sazuka et al., 1996, 1997).

TEA AND DENTAL HEALTH

Epidemiological surveys have reported that a considerable percentage of populations, who drink tea on a regular basis, have a reduced number of carious teeth (Ramsey, 1975; Onisi, 1985; Cao, 1987). Green tea and various catechins are also known to inhibit the growth of cariogenic bacteria by preventing the adherence and growth of such bacteria on the dental surface (Otake, 1991; Sakanaka, 1990). Also, the adsorption of theaflavins to dental enamel may play an important role in curbing the formation of plaque, and keeping the teeth clean and shining.

A recent study on the bio-availability of fluoride (from tea) demonstrated that 34% of the fluoride was retained in the oral cavity and that some of this showed a strong binding ability to the tooth enamel, thereby indicating that tea could be effective towards replenishment of fluoride to the oral cavity, as well prevent dental decay and gum diseases (Simpson, 2001).

It has also been reported that the tannins in tea can inhibit salivary amylase, thus reducing the carcinogenic potential of starch-containing foods (Zhang, 1998). A number of studies have also demonstrated the effectiveness of tannic acid to combat the growth of S mutans bacteria (Iizuka, 1980; Paolino, 1980; Wu-Yuan, 1988; Tachibana, 1989; Itoh, 1991; Yu et al., 1995), a major cause for the development of dental plaque. Tea tannins, along with catechin, caffeine and tocopherol can be effective in increasing the acid resistance of tooth enamel (Iizuka, 1980).

TEA AND FLUID REPLENISHMENT

In the early days, tea was not considered to be an alternative to fluid replacement because of its diuretic action (due to the presence of caffeine). However, it was later observed that tea did not exert any diuretic action, unless the amount of tea consumption exceeded 6 cups (i.e. 250-300 mg caffeine equivalent) (Nussberger, 1990; Neuhäuser-Berthold, 1997; Martof and Knox, 1997; Stookey, 1999; Passmore, 1987; Grandjean, 2000).

It may be mentioned that owing to the volume of fluid that is consumed with tea, the habit of tea consumption significantly contribute towards the daily recommended fluid intake in adults, thereby improving the hydration status (Todorovic and Micklewright, 1997).

GASTROINTESTINAL SYSTEM

Administration of black tea extract, to rats, for seven days inhibited the development of both aspirin and cold-restrained stress induced acute gastric ulcer. Tea extract was found to reduce acid and peptic activity of gastric secretion, induced by aspirin and cold-restrained stress (Maity et al.,

1995). Moreover, tea extract increased gastric glutathione peroxidase, hexosamine and sialic acid contents, thereby suggesting that tea probably stimulate the overall protection of gastric mucosa (Maity et al., 1998).

Hot water extract of black tea significantly accelerated the gastro-intestinal transit (*in vivo*) in mice. This effect was probably due to the influence of tea on prostaglandin and nitric oxide generation (Maity et al., 2001; Chaudhuri et al., 2000).

From studies in our laboratory, we observed that ingestion of black tea infusion (Darjeeling variety) significantly inhibited aspirin, alcohol and serotonin-induced ulceration in rats (Tables 4.1-4.3). It was also observed that following black tea (infusion) pretreatment there was significant restoration of superoxide dismutase (SOD) and catalase (CAT) enzyme activity in gastric and liver tissue following ethanol-induced oxidative stress (Nagchaudhuri et al., 2004).

In some other studies it was observed that tea infusion exhibited potent antidiarrhoeal activity, probably related to the presence of tannins (Besra et al., 2003).

Table 4.1 Effect of black tea on aspirin-induced gastric ulceration

(Mean±standard error; n=6)

Drug	*Dose (mg/kg)*	*Lesion index (mm)*	*Inhibition %*	*P value*
Control	–	28.66±2.836	–	–
Black tea	200	9.3±0.881	67.55	<0.01
Black tea	1000	2.5±0.844	91.27	<0.001

Table 4.2 Effect of black tea on alcohol-induced gastric ulceration

(Mean±standard error; n=6)

Drug	*Dose (mg/kg)*	*Lesion index (mm)*	*Inhibition %*	*P value*
Control	–	31.75±6.776	–	–
Black tea	200	2.769±1.83	91.28	<0.02
Black tea	1000	6.349±2.35	80.00	<0.01

Table 4.3 Effect of black tea on serotonin-induced gastric ulceration

(Mean±standard error; n=6)

Drug	*Dose (mg/kg)*	*Lesion index (mm)*	*Inhibition %*	*P value*
Control	--	13.5±1.384	----	--
Black tea	200	8.66±0.714	36.29	<0.02
Black tea	1000	3.33±0.614	75.55	<0.01

NEUROLOGICAL EFFECTS OF TEA

Both green and black tea extract significantly potentiated the onset, duration and mortality in experimentally induced convulsion in mice. These extracts are suggested to act probably by influencing the calcium channels (Gomes et al., 1999). Tea infusion has been found to improve alertness, which could be attributed to both caffeine and theanine (Hindmach et al., 1998; Juneja et al., 1999).

ANTIHYPERGLYCEMIC ACTIVITY

The hot water extract of black tea was found to significantly reduce the blood glucose level of streptozotocin-induced diabetic rats. Studies have also shown that the constituents of tea may indirectly stimulate the secretion of insulin (Gomes et al., 1995). It has also been observed that green tea can reduce glucose transportation from intestine (Shimizu, 1999).

ANTIINFLAMMATORY AND ANTIARTHRITIC ACTIONS

Studies have shown the effectiveness of tea polyphenols in reducing the release of histamine and also hyaluronidase activity, during the inflammatory process (Alexis et al., 1999; Toyoda et al., 1997). Such anti-inflammatory and antiarthritic activity (induced with collagen) has also been reported by Middleton (1998). The saponins present in tea leaf were found to inhibit carrageenin-induced rat paw oedema in a dose-dependent manner; probably by inhibition of the hyaluronidase enzyme activity and also antagonizing the action of leukotriene D_4 (Sagesaka et al., 1996). Tea has also been found to be useful in rheumatoid arthritis (Haqqi et al., 1999). According to clinical reports (from a study conducted in the USA), tea consumption was related to increased bone mineral density and relief from osteoporosis particularly in women (Chen et al., 2003).

From study conducted in our laboratory, it was observed that chronic treatment with black tea infusion (200 and 1000 mg/kg; oral) significantly inhibited carrageenin-induced paw oedema and formaldehyde-induced chronic polyarthritis in albino rats (Figs. 4.3 and 4.4). Such antiarthritic activity was further supported by our observation from another study, where it was found that the tea infusion significantly inhibited depolymerization (superoxide-induced) of hyaluronic acid.

Similarly in other studies related to enzymatic (NADPH dependent) and non-enzymatic ($FeCl_3$/Ascorbic and Cu^{2+}/ascorbate-induced) lipid

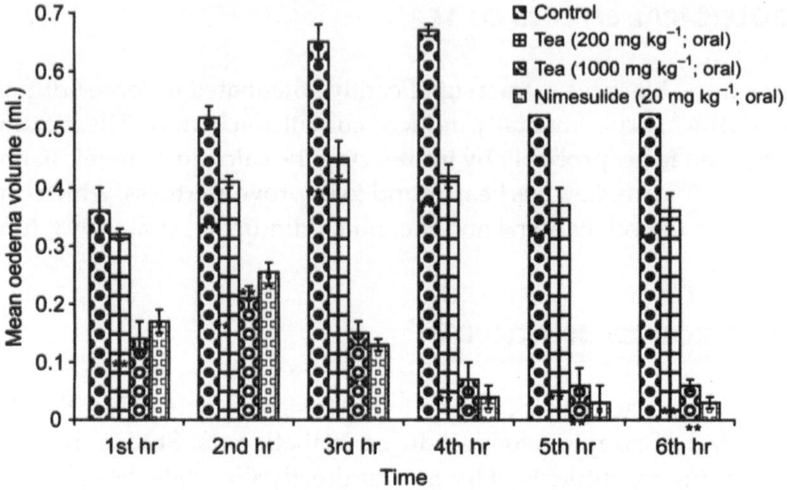

Fig. 4.3 Effect of black tea on carrageenin-induced paw oedema

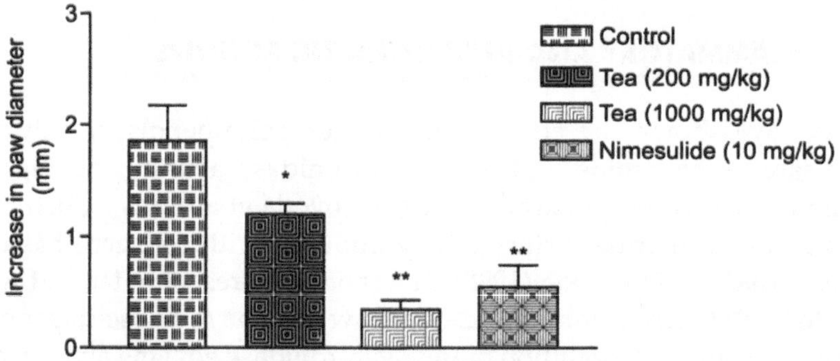

Fig. 4.4 Effect of black tea on formaldehyde-induced arthritis

peroxidation, black tea infusion (Darjeeling variety) significantly inhibited malondialdehyde formation in a dose-dependent manner. Such antipre-oxidative (Figs. 4.5 and 4.6) activity was comparable to that of standard drugs Vitamin C, E, sodium metabisulphite and quercetin (Nagchaudhuri et al., 2004).

SKIN AND TOPICAL WOUND HEALING ACTIVITY

Tea leaves can inhibit bleeding from cuts and wounds, probably due to the involvement of tannins (Stephen et al., 2003), when applied topically. Tea polyphenols (particularly ECGC and ECG) have been found to protect skin

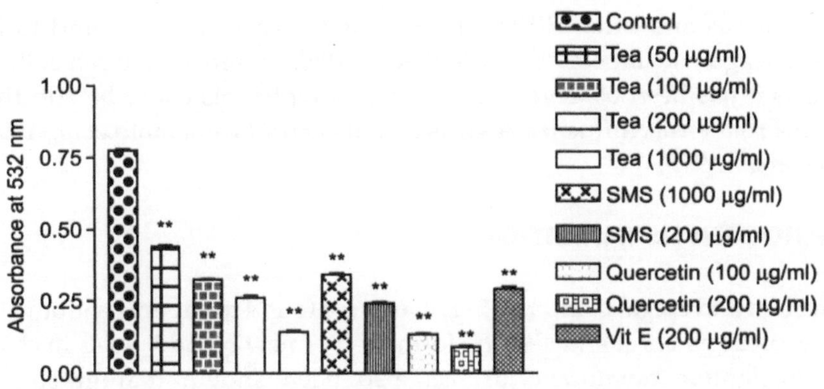

Fig. 4.5 Effect of black tea on NADPH dependent lipid peroxidation (enzymatic)

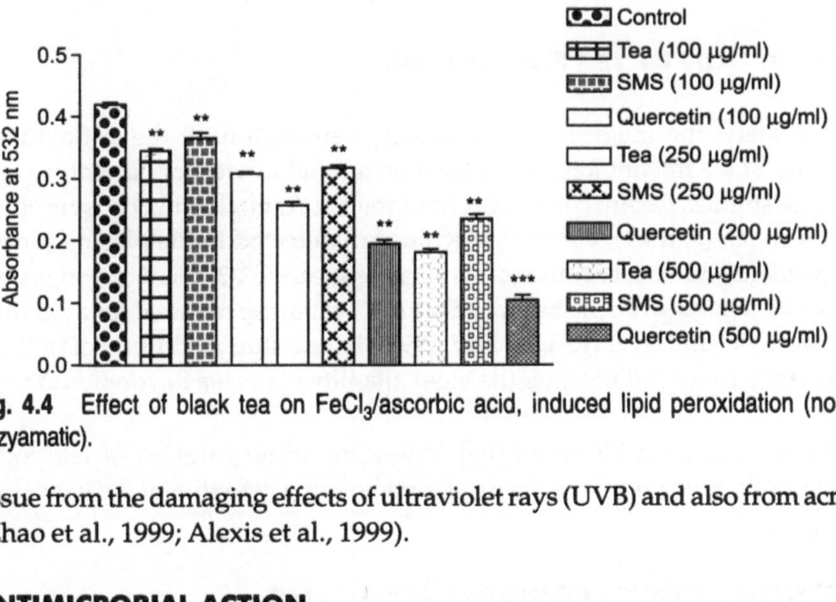

Fig. 4.4 Effect of black tea on $FeCl_3$/ascorbic acid, induced lipid peroxidation (non-enzyamatic).

tissue from the damaging effects of ultraviolet rays (UVB) and also from acne (Zhao et al., 1999; Alexis et al., 1999).

ANTIMICROBIAL ACTION

Tea infusion has been found to be effective against bacteria (particularly gram-positive). Tea catechins demonstrate strong antimicrobial activity against *Helicobacter pylori*, with a MIC of 8 mcg per ml (Diker, 1994). In *Helicobacter* infected Mongolian gerbils, tea catechins eradicated infection upto 10-36%, with a significant decrease in mucosal hemorrhage and erosion. Studies have shown that tea extracts show inhibitory effects against *Salmonella typhi, Campylobacter jejuni, Campylobacter coli* (Diker, 1993), *Helicobacter pylori* (Diker, 1994; Yanagawa, 2003), *Shigella, Clostridium,*

Pseudomonas and other different organisms. Tea was also found to be effective against amoebiasis (Ghosal et al., 1993). From animal cell culture studies it has been observed that the tea polyphenols could be effective against rota virus, influenza A virus and also HIV (Yamamoto et al., 1997; Pillai et al., 1999).

ANGIOGENESIS INHIBITION

Green tea and epigallocatechin-3-gallate were found to suppres endothelial cell growth (*in vitro*) and also the formation of new blood vessels in chick chorioallantoic membrane. It has also been shown that green tea consumption prevented corneal neo-vascularisation induced with an angiogenic factor (Yihai, 1999).

ABSORPTION OF TEA FLAVONOIDS

Until lately, the majority of researches, demonstrating the antioxidant activity of tea flavonoids, were based on animal models or laboratory cell culture studies. Recent clinical studies indicate that following consumption of green tea (extract), various catechins are detected in the plasma (dose-dependent), in concentrations varying between 0.2-2.0% of the ingested amount; it has also been observed that 1.4-2.4 hrs. are required for attaining the peak plasma level (van het Hof, 1998). In one study, addition of milk to tea preparations did not alter the bioavailabilty of the tea flavonoids (Yang, 1999).

It has also been observed that long-term administration of tea polyphenols (in rats) improves the drug metabolizing activities of the liver (Liu et al., 2003).

POTENTIAL HEALTH CONCERNS ASSOCIATED WITH TEA CONSUMPTION

Tea and Iron Absorption

Individual requirements for dietary iron vary according to age, gender and diet. Low iron intake is common amongst the British population, especially among females (Gregory, 1990).

The inhibitory effects of tea drinking on iron absorption was first identified in a study that used to test meals fed under experimental conditions (Disler, 1975). A few studies have attempted to quantify the effect of tea on non-haem iron absorption (Reddy, 2000; Hallberg et al., 2000; Zijp, 2000). These studies concluded that tea possesses an inhibiting effect on such iron absorption,

Studies on tea consumption in children claimed a higher incidence of anaemia amongst tea drinkers (Merhav et al., 1985; Wilson et al., 1999; Gibson, 1999). However, there is no evidence to suggest that tea drinking should be restricted in healthy individuals who are consuming a well-balanced, mixed diet.

In vegetarians, the absorption of iron (non-heam form) is restricted, thus this section of people are advised to avoid tea consumption with major meals, as the iron in the food is known to be chelated by tea.

Tea leaves contain aluminium hydroxide, which has raised the concern about the possible link between tea consumption and Alzheimer's disease (Flaten and Odegard, 1988). At this moment there is insufficient scientific data to establish such relationship, since much of the aluminium remain bound to organic species such as theaflavins, thearubigins and other polyphenols, which are not easily absorbed from the gastrointestinal tract.

CONCLUSION

On the basis of available scientific information, it is evident that modern science has further reinforced the age-old perception about the health benefits associated to tea consumption. At the present moment, science has definitely revealed the extent of health benefits associated with tea and the possible mechanisms related to its pharmacological activities. It is pertinent to mention that there is an absolute necessity to promote an in-depth study on the biological activities (including clinical trials) and mechanism of action related to tea consumption because focusing on the benefits associated with tea (disease prevention) could help us to save not only thousands of lives but may also have a positive impact on the health budget of the nation.

Considering the amount of research data that has been generated on tea (Black and Green) and its polyphenols, during the past few years, it is now time for us to realize the immense beneficial potential of tea as a beverage, thus once again reinforcing the ancient reports on the healing powers of tea.

Acknowledgements

The authors thankfully acknowledge the support of M/s. S. Karmakar, K.T. Mani Senthil Kumar and D. Adhikari in preparation of this manuscript.

REFERENCES

Alexis, A.F., V.A. Jones and M.J. Stiller. 1999. Potential therapeutic applications of tea in dermatology. Int. J. Dermatol. 38: 735-43.

Arts, I., P.C. Hollman, H.B. Bueno De Mesquita, E.J. Feskens and D. Kromhout. 2001. Dietary catechins and epithelial cancer incidence: the Zutphen elderly study. Int. J. Cancer. 92: 298-302.

Balentine, D.A., S.A. Wiseman and L.C. Bouwens. 1997. The chemistry of tea flavonoids. Crit. Rev. Food Sci. Nutr. 37: 693-704.

Besra, S.E., A. Gomes, D.K. Ganguly and J.R. Vedasiromoni. 2003. Antidiarrhoeal activity of hot water extract of black tea *Camellia sinensis*. Phytother. Res. 17: 380-4.

Bushman, J.L. 1998. Green tea and cancer in humans: a review of the literature. Nutr. Cancer 31: 151-9.

Cao, J. 1987. Observation of caries incidence among a tea-drinking population. J. Dental Health 31: 86-9.

Cao, Y. and R. Cao. 1999. Angiogenasis inhibited by drinking tea. Nature 398: 381.

Chaudhuri, L., S. Basu, P. Seth, T. Chaudhuri, S.E. Besra, J.R. Vedasiromoni and D.K. Ganguly. 2000. Prokinetic effect of black tea on gastrointestinal motility. Life Sci. 21: 847-54.

Chen, Y.C., Y.C. Liang, S.Y. Lin-Shiau, C.T. Ho and J.K. Lin. 1999. Inhibition of TPA-induced protein kinase C and transcription activator protein-1 binding activities by theaflavin-3,39-digallate from black tea NIH3T3 cells. J. Agric. Food Chem. 47: 1416-21.

Chen, Z., M.B. Pettinger, C. Ritenbaugh, A.Z. LaCroix, J. Robbins, B.J. Caan, D.H. Barad and I.A. Hakim. 2003. Habitual tea consumption and risk of osteoporosis: a prospective study in the women's health initiative observational cohort. Am. J. Epidemiol. 15: 772-81.

Corvazier, E. and J. Maclouf. 1985. Interference of some flavonoids and non-steroidal anti-inflammatory drugs with oxidative metabolism of arachidonic acid by human platelets and neutraphils. Biochem. Biophys. Acta. 835: 315-321.

De-Whalley, C.V., S.M. Rankin, J.R. Hoult, W. Jessup and D. S. Leake. 1990. Flavonoids inhibit the oxidative modification of Low Density Lipoproteins by macrophages. Biochem. Pharmacol. 39: 1743-1750.

Diker, K.S. 1994. The bactericidal activity of tea against *Helicobacter pylori*. Lett. Appl. Microbiol. 19: 299-300.

Diker, K.S. 1993. The bactericidal activity of tea against *Campylobacter jejuni* and *Campylobacter coli*. Lett. Appl. Microbiol. 12: 34-5.

Diker, K.S. 1994. The bactericidal activity of tea against *Helicobacter pylori*. Lett. Appl. Microbiol. 19: 299-300.

Disler, P.B. 1975. The effect of tea on iron absorption. Gut. 16: 193-200.

Dreosti, I.E. 1997. Inhibition of carcinogenesis by tea: the evidence from experimental studies. Crit. Rev. Food Sc. Nutr. 37: 761-70.

Du Toit, R. 2001. Comparison of the antioxidant content of fruits, vegetables and teas measured as Vitamin C equivalents. Toxicol. 166: 63-9.

Duffy, S.J. 2001. Short- and long-term black tea consumption reverses endothelial dysfunction in patients with coronary artery disease. Circulation 10: 151.

Dufresne, J. and Edward R. Farnworth. 2001. A review of latest research findings on the health promotion properties of tea. J. Nutr. Biochem. 12: 404-421.

Flaten, T.P. and M. Odegard. 1988. Tea, aluminium and Alzheimer's disease. Food. Chem. Toxicol. 26: 959-60.

Fujiki, H., M. Suganuma, S. Okabe, E. Sueoka, K. Suga, K. Imai, K. Nakachi and S. Kimura. 1999. Mechanistic findings of green tea as cancer preventive for humans. Proc. Soc. Exper. Biol. Med. 220: 225-8.

Geleijnse, J.M., L.J. Launer, A. Hofman, H.A. Pols and J.C. Witteman 1999. Tea flavonoids may protect against atherosclerosis: the Rotterdam study. Arch. Intern. Med. 159: 2170-4.

Ghosal, S., B.N. Krishna Prasad, K. Raj. and A.P. Bhadur. 1993. Antiamoebic and antibacterial properties of black tea. Tea Science and Human Health: Proceedings of the International Symposium TEA TECH 1993, Tea Research Association, Kolkata. pp. 119-124.

Gibson, S.A. 1999. Iron intake and iron status of pre-school children: associations with breakfast cereals, vitamin C and meat. Pub. Health Nutr. 2: 521-8.

Gomes, A., J.R. Vedasiromoni, M. Das, R.M. Sharma. and D.K. Ganguly. 1995. Anti-hyperglycemic effect of black tea *Camellia Sinensis* in rat J. Ethnopharmacol. 45: 223-2.

Gomes, A., M. Das, J.R. Vedasiromoni and D.K. Ganguly. 1999. Proconvulsive effect of tea *Camellia sinensis* in mice. Phyt. Res. 13: 376-9.

Grandjean, A.C. 2000. The effect of caffeinated, non-caffeinated, caloric and non-caloric beverages on hydration. J. Am. Coll. Nutr. 19: 591-600.

Gregory, J. 1990. The Dietary and Nutritional Survey of British Adults.London: HMSO.

Gutman, R.L. and B.H. Ryu. 1996. Rediscovering tea. An exploration of the scientific literature, Herbal Gram. 37: 33–48.

Hallberg, L. and L. Hulthen. 2000. Prediction of dietary absorption: an algorithm for calculating absorption and bioavailability of dietary iron. Am. J. Clin. Nutr. 71: 1147-60.

Haqqi, T.M., D.D. Anthony, S. Gupta, N. Ahmad, M.S. Lee, G.K. Kumar and H. Mukhtar. 1999. Prevention of collagen-induced arthritis in mice by a polyphenolic fraction from green tea. Proc. Natl. Acad. Sci. 96: 4524-9.

Hertog, M.G., D. Kromhout, C. Aravanis, H. Blackburn, R. Buzina, F. Fidanza, S. Giampaoli, A. Jansen, A. Menotti and S. Nedeljkovic. 1995. Flavonoid intake and long-term risk of coronary heart disease and cancer in the seven countries study. Arch. Int. Med. 155: 381.

Hertog, M.G., P.M. Sweetnam, A.M. Fehily, P.C. Elwood and D. Kromhout. 1997. Antioxidant flavonols and ischaemic heart disease in a Welsh population of men: the Caerphilly study. Am. J. Clin. Nutr. 65: 1489-94.

Hertog, M.G.L. 1993. Dietary antioxidants flavonoids and risk of coronary heart disease: the Zutphen Elderly Study. Lancet. 155: 381

Hindmarch, I., P.T. Quinlan, K.L. Moore and C. Parkin. 1998. The effects of black tea and other beverages on aspects of cognition and psychomotor performance. Psychopharmacol. 139: 230-8.

Hofbauer, R., M. Frass, B. Gmeiner, S. Handler, W. Speiser and S. Kapiotis. 1999. The green tea extract epigallocatechin is able to reduce neutrophil transmigration through monolayers of endothelial cells. Wien. Klin. Woche. 111: 278-82.

Iizuka, S. 1980. Inhibitory effect of tannic acid on the lactate metabolism in human saliva. Dentistry, 67: 770-778.

Inoue, M. 2001. Regular consumption of green tea and the risk of breast cancer recurrence: follow-up study from the Hospital-based Epidemiologic Research Program at Aichi Cancer Center HERPACC. Jap. Cancer Letters: 167: 175-82.

Inoue, M., K. Tajima, K. Hirose, N. Hamajima, T. Takezaki, T. Kuroishi and S. Tominaga. 1998. Tea and coffee consumption and the risk of digestive tract cancers: data from a comparative case-reference study in Japan. Can. Causes Cont. 9: 209-16.

Itoh, T. 1991. Inhibition of cariogenic factor of mutans streptococci by tannic acid. Nihon Univ. J. Oral Sci. 17: 115-126.

Jakun, J. 1997. Why drinking green tea could prevent cancer. Nature 387: 561.

Ji, B.T., W.H. Chow, G. Yang, J.K. McLaughlin, R.N. Gao, W. Zheng, X.O. Shu, F. Jin, J.F. Fraumeni Jr and Y.T. Gao. 1996. The influence of cigarette smoking, alcohol and green tea consumption on the risk of carcinoma of the cardia and distal stomach in Shanghai, China. Cancer 77: 2449-57.

Juneja, L.R., D.C. Chu, T. Okubo, Y. Nagato and H. Yokogoshi. 1999. L-theanine–a unique amino acid of green tea and its effect in humans. Trends Food Sci. Technol. 10: 199-204.

Katiyar, S.K. 1996. Tea in chemoprevention of cancer: epidemiological and experimental studies. Review. Int. J. Oncol. 5: 425-38.

Keli, S.O. 1996. Dietary flavonoids, antioxidant vitamins, and incidence of stroke: the Zutphen study. Arch. Intern. Med. 156: 637-642.

Klatsky, A.L., G.D. Friedman, M.A. Armstrong and A.L. Klatsky. 1990. Coffee use prior to myocardial infarction restudied: heavier intake may increase the risk. Am. J. Epidemiol. 132: 479-88.

Knekt, P., R. Jarvinen, A. Reunanen and J. Maatela. 1996. Flavonoid intake and coronary mortality in Finland: a cohort study. Brit. Med. J. 312: 478-81.

Knekt, P., R. Jarvinen, R. Seppanen, M. Hellovaara, L. Teppo, E. Pukkala and A. Aromaa. 1997. Dietary flavonoids and the risk of lung cancer and other malignant neoplasms. Am. J. Epidem. 146: 223-30.

Kono, S. 1988. A case-control study of gastric cancer and diet in Northern Kyushu, Japan. Jpn. J. Cancer Res. 79: 1067-74.

Lin, Y.L. 1998. Hypolipidemic effect of green tea leaves through induction of antioxidant and Phase II enzymes including superoxidase dismutase, catalase, and glutathione S-transferase in rats. J. Agric. Food. Chem. 46: 1893-9.

Liu, Tao-Tao, N. Liang, L. Yan, F. Yang, Y. Lu, Zi. Meng and L. Zhang 2003. Effects of long-term tea polyphenols consumption on hepatic microsomal drug-metabolizing enzymes and liver function in Wistar rats. World J. Gastroenterol. 9: 2742-2744.

Lyn-Cook, B.D., T. Rogers, Y. Yan, E.B. Blann, F.F. and G. J. Kadlubar Hammons. 1999. Chemopreventive effects of tea extracts and various components on human and pancreatic and prostate tumor cells in vitro. Nutr. Cancer 35: 80-86.

Maity, S., J.R. Vedasiromoni and D.K. Ganguly. 1995. Antiulcer effect of hot water extract of black tea. J. Ethnopharmacol. 46: 167-174.

Maity, S., J.R. Vedasiromoni and D.K. Ganguly. 1998. Role of glutathione in the antiulcer effect of hot water extract of black tea *Camellia sinensis*. Jpn. J. Pharmacol. 78: 285-92.

Maity, S., J.R. Vedasiromoni, L. Chaudhuri and D.K. Ganguly. 2001. Role of reduced glutathione and nitric oxide in the black tea extract-mediated protection against ulcerogen-induced changes in motility and gastric emptying in rats. Jpn. J. Pharmacol. 85: 358-64.

Martof, M.T. and D.K. Knox. 1997. The effect of xanthines on fluid balance. Clin. Nurs. Res. 6: 186-196.

Merhav, H., Y. Amitai, H. Palti and S. Godfrey. 1985. Tea drinking and microcytic anaemia in infants. Am. J. Clin. Nutr. 41: 1210-3.

Middleton, E. 1994. The impact of plant flavonoids on mammalian biology: implications for immunity, inflammation and cancer. In: Harborne JB, editor. The flavonoids: advances in research since 1986. London: Chapman and Hall, pp. 619-52.

Middleton, E. 1998. Effect of plant flavonoids on immune and inflammatory cell function. Adv. Exp. Med Biol. 439: 175-82.

Naasani, I., H. Seimiya and T. Tsuruo. 1998. Elomerase inhibition, telomerase shortening, and senescence of cancer cells by tea catechin. Biochem. Biophys. Res. Comm. 249: 391-6.

Nagchauduri, A.K., T. Sen and D.K. Roy. 2004. Phramacological evaluation of the medicinal properties of Indian black tea. Ann. Sc. Report. 2003-2004. National Tea Research Foundation, Kolkata.

Neuhäuser-Berthold, M. 1997. Coffee consumption and total body water homeostasis as measured by fluid balance and bioelectrical impedance analysis. Ann. Nutr. & Metabol. 41: 29-36.

Nussberger, J. 1990. Caffeine-induced diuresis and atrial natriuretic peptides. J. Card. Pharmacol. 15: 685-691.

Onisi, M. 1985. Analysis of data obtained from 5 years tea drinking program for the caries prevention by means of the linear caries extent/risk relation. J. Dental Health. 35: 138-9.

Osada, K., M. Takahashi, S. Hoshina, M. Nakamura, S. Nakamura and M. Sugano. 2001. Tea catechins inhibit cholesterol oxidation accompanying oxidation of low-density lipoprotein in vitro. Comp. Biochem. Physiol. & Toxicol. Pharmacol. 128: 153-64.

Osada, K., M. Takahashi, S. Hoshina, M. Nakamura, S. Nakamura and M. Sugano. 2001. Tea catechins inhibit cholesterol oxidation accompanying oxidation of low-density lipoprotein in vitro. Comp. Biochem. Physiol. C Toxicol. Pharmacol. 128: 153-64.

Otake, S. 1991. Anticaries effects of polyphenolic compounds from Japanese green tea. Caries Res. 25: 438-43.

Paganga, G., N. Miller and C.A. Rice-Evans. 1999. The polyphenolic content of fruit and vegetables and their antioxidant activities. What does a serving constitute? Free Rad. Res. 30: 153-162.

Paolino, V.J. 1980. Inhibition of dextran synthesis by tannic acid. Am. Assoc. Dent. Res.: 488 abstract.

Passmore, A.P. 1987. Renal and cardiovascular effects of caffeine: dose response study. Clin. Sci. 72: 749-56.

Pearson, D.A. 1998. Inhibition of endothelial cell-mediated low-density lipoprotein oxidation by green tea extracts. J. Agric. Food. Chem. 46: 1445-9.

Peters, U., C. Poole and L. Arab. 2001. Does tea affect Cardiovascular Disease? A Meta-Analysis. Am. J. Epidemiol. 154: 6: 495-501.

Pillai, S.P., L.A. Mitscher, S.R. Menon, C.A. Pillai and D.M. Shankel. 1999. Antimutagenic/antioxidant activity of green tea components and related compounds, J. Envir. Pathol. Toxicol. Oncol. 18: 147-58.

Pillai, S.P., L.A. Mitscher, S.R. Menon, C.A. Pillai and D.M. Shankel. 1999. Antimutagenic/antioxidant activity of green tea components and related compounds. J. Envir. Pathol. Toxicol. Oncol. 18: 147-58.

Polette, A. 1996, N-3 fatty acid-induced lipid peroxidation in human platelets is prevented by catechins. Thrombosis and Haemostasis 95: 945-9.

Ramsey, A.C. 1975. Fluoride intakes and caries increments in relation to tea consumption by British children. Caries Res. 9: 312.

Reddy, M.B. 2000. Estimation of non-haem iron bioavailability from meal composition. Amer. J. Clin. Nutr. 71: 937-430.

Rimm, E.B., M.B. Katan, A. Ascherio, M.J. Stampfer and W.C. Willett. 1996. Relation between intake of flavonoids and risk for coronary heart disease in male health professionals. Ann. Intern. Med. 125: 384-9.

Sagesaka, Y.M., T. Uemura, Y. Suzuki, T. Sugiura, M. Yoshida, K. Yaamaguchi and K. Kyuki 1996. Antimicrobial and anti-inflammatory actions of tea-leaf. Yakugaku Zasshi 116: 238-43.

Sakanaka, S. 1990. Inhibitory effects of green tea polyphenols on glucan synthesis and cellular adherence of cariogenic Streptococci. Agric. Biol. Chem. 54: 2925-9.

Sazuka, M., H. Imazawa, Y. Shoji, T. Mita, Y. Hara and M. Isemura. 1997. Inhibition of collagenases from mouse lung carcinoma cells by green tea catechins and black tea theaflavins. Biosci. Biotech. Biochem. 61: 1514-6.

Sazuka, M., T. Itoi, Y. Suzuki, S. Odani, T. Koide and M. Isemura. 1996. Evidence for the interaction between (2)-epigallocatechin gallate and human plasma protein fibronectin, fibrinogen, and histidine-rich glycoprotein. Biosci. Biotech. Biochem. 60: 1317-9.

Sesso, H.D., J.M.Gaziano, J.E. Buring and C.H. Hennekens. 1999. Coffee and tea intake and the risk of myocardial infarction. Am. J. Epidemiol. 149: 162-167.

Setiawan, V.W., Z.F. Zhang, G.P. Yu, Q.Y. Lu, Y.L. Li, M.L. Lu, M.R. Wang, C.H. Guo, S.Z. Yu, R.C. Kurtz and C.C. Hsieh. 2001. Protective effect of green tea on the risks of chronic gastritis and stomach cancer. Int. J. Cancer 92: 600-604.

Shimizu, M. 1999. Modulation of the intestinal function by food substances. Nahrung. 43: 154-8.

Simpson, A. 2001. The bioavailability of fluoride from black tea. J. Dentistry. 29: 15-21.

Smith, D.M. 2001. Green tea induces polyphenols epigallocatechin inhibts DNA replication and consequently induces leukaemia cell apoptosis. Int. J. Mol. Med. 76: 645-52.

Stensvold, I. 1992. Tea consumption. Relationship to cholesterol, blood pressure and coronary and total mortality. Preventative Med. 21: 546.

Stephen, H., W. Bollag, J. Lewis, Q. Huang, B. Singh, M. Sharawy, T. Yamamoto and G. Schuster. 2003. Green Tea Polyphenols Induce Differentiation and Proliferation in Epidermal Keratinocytes. J. P. E. T. 306: 29-34.

Stookey, J.D. 1999. The diuretic effects of alcohol and caffeine and total water intake misclassification. Eur. J. Epidemiol. 15: 181-188.

Tachibana, Y. 1989. The effect of suppression of the tannin fluoride preparation upon the development of dental plaque. J. Jpn. Endodont. 10: 65.

Tijburg L.B., T. Mattern, J.D. Folts, U.M. Weisgerber and M.B. Katan. 1997. Tea flavonoids and cardiovascular diseases – a review. Crit. Rev. Food. Sci. Nutr. 37: 771-85.

Todorovic and Micklewright. 1997. A pocket guide to clinical nutrition, 2 edn.

Parenteral and Enteral Nutrition (PEN) Group of the British Dietetic Association. Birmingham: British Dietetic Association.

Toyoda, M., K. Tanaka, K. Hoshino, H. Akiyama, A. Tanimura and Y. Saito. 1997. Profiles of potentially antiallergic flavonoids in 27 kinds of health tea and green tea infusions. J. Agric. Food Chem. 45: 2561-4.

Trevisanto, S. and Y. Kim. 2000. Tea and health. Nutr. Rev. 58: 1-10.

Tzeng, S.H., W.C. Ko, F.N. Ko and C.M. Teng. 1991. Inhibition of platelet aggregation by some flavonoids. Thromb. Res. 64: 91-100.

van het Hof, K. 1998. Bioavailability of catechins from tea: the effect of milk. Eur. J. Clin. Nut. 52: 356.

Vinson, J. A. and Y.A. Dabbagh. 1998. Effect of green and black tea supplementation on lipids, lipid oxidation and fibrinogen in hamster: mechanisms for the epidemiological benefits of tea drinking. FEBS Lett. 433: 44-6.

Vinson, J.A. 1995. Plant flavonoids, especially tea flavonols, are powerful antioxidants using an in vitro oxidation model for heart disease. J. Agric. Food. Chem. 43: 2800-2802.

Wang, Z.Y., J.Y. Hong, M.T. Huang, K.R. Reuhl, A.H. Conney and C.S. Yang. 1992. Inhibition of N-nitrosodiethylamine- and 4-methylnitrosamino.-1-3-pyridyl.-1-butanone-induced tumorigenesis in A/J mice by green tea and black tea. Cancer Res. 52: 1943-7.

Wang, Z.Y., S.J. Cheng, Z.C. Zhou, M. Athar, W.A. Khan, D.R. Bickers and H. Mukhtar. 1989. Antimutagenic activity of green tea polyphenols. Mutat. Res. 223: 273-285.

Wilson, C., C.C. Grant and C.R. Wall. 1999. Iron deficiency anaemia and adverse dietary habits in hospitalized children. N. Z. Med. J. 112: 203-6.

Wisburger, J.H. 1999. Tea and health: the underlying mechanisms. Proc. Soc. Exper. Biol. Med. 220: 271-5.

Wu-Yuan, C.D. 1988. Gallotannins inhibit growth, water insoluble glucan synthesis and aggregation of mutans streptococci. J. Dent. Res. 67: 51-55.

Xu, Y., C.T. Ho, S.G. Amin, C. Han and F.L. Chung 1992. Inhibition of tobacco-specific nitrosamine-induced lung tumorigenesis in A/J mice by green tea and its major polyphenols as antioxidants. Cancer Res. 52: 3875-9.

Yamamoto, T., L.R. Juneja, D.C. Chu and M. Kim. 1997. Chemistry and Applications of Green Tea. CRC Press LLC: Boca Raton, USA.

Yanagawa, Y. 2003. A combination effect of epigallocatechin gallate, a major compound of green tea catechins, with antibiotics on *Helicobacter pylori* growth in vitro. Curr. Microbiol. 47: 244-9.

Yang, C.S. 1999. Inhibition of carcinogenesis by tea: bioavailability of tea polyphenols and mechanisms of actions. Proc. Soc. Exper. Biol. Med. 220: 213-7.

Yang, C.S. and Z.Y. Wang. 1993. Tea and cancer. J. Natl. Cancer Inst. 85: 1038-49.

Yang, T.T. and M.W. Koo. 1997. Hypocholesterolemic effects of Chinese tea. Pharmacol. Res. 35: 505-12.

Yang, C.S., and Z.Y. Wang 1993. Tea and cancer. J. Natl. Cancer Inst. 85: 1038-49.

Yochum, L., L.H. Kushi, K. Meyer and A.R. Folsom. 1999. Dietary flavonoid intake and risk of cardiovascular disease in postmenopausal women. Am. J. Epidemiol. 149: 943-9.

Yokozawa, T., E. Dong, T. Nakagawa, H. Kashiwagi, H. Nakagawa, S. Takeuchi and H.Y. Chung. 1998. *In vitro* and *in vivo* studies on the radical scavenging activity of tea. J. Agric. Food Chem. 46: 2143–51.

Yokozawa, T., H. Oura, S. Sakanaka, S. Ishigaki and M. Kim. 1994. Depressor effect of tannin in green tea on rats with renal hypertension. Biosci. Biotech. Biochem. 58: 855-8.

Yokozawa, T., H.Y. Chung, L.Q. He and H. Oura. 1996. Effectiveness of green tea tannin on rats with chronic renal failure. Biosci. Biotech. Biochem. 60: 1000-5.

Yoshida, H., T. Ishikawa, H. Hosoai, M. Suzukawa, M. Ayaori, T. Hisada, S. Sawada, A. Yonemura, K. Higashi, T. Ito, K. Nakajima, T. Yamashita, K. Tomiyasu, M. Nishiwaki, F. Ohsuzu and H. Nakamura. 1999. Inhibitory effect of tea flavonoids on the ability of cells to oxidise low-density lipoprotein. Biochem. Pharmacol. 58: 1695-703.

Yu, G.P. and C.C. Hsieh. 1991. Risk factors for stomach cancer: a population based case-control study in Shanghai. Can. Causes Contr. 2: 169-74.

Yu, G.P., C.C. Hsieh, L.Y. Wang, S.Z. Yu, X.L. Li and T.H. Jin. 1995. Green tea consumption and risk of stomach cancer: a population based case-control study in Shanghai. Can. Causes Contr. 6: 532-8.

Yu, H., T. Oho and L.X. Xu 1995. Effects of several tea components on acid resistance of human tooth enamel. J. Dent. 23: 101-105.

Zhang, J. 1998. Inhibition of salivary amylase by black and green tea and their effects on the intraoral hydrolysis of starch. Caries Res. 32: 233-8.

Zhao, J., X.E.Y. Jin, Z.S. Zheng, Y.J. Zhang, M. Athar, V.A. DeLeo, H. Mukhtar, D.R. Bickers and Z.Y. Wang 1999. Photoprotective effect of black tea extracts against UVB-induced phototoxicity in skin. Phtotchem. Photobiol. 70: 637-44.

Zhu, Q.Y., Y. Huang and Z.Y. Chen. 2000. Interaction between flavonoids and alpha-tocopherol in human low-density lipoprotein. J. Nutr. Biochem. 11: 14-21.

Zijp, I.M. 2000. Effect of tea and other dietary factors on iron absorption. Crit. Rev. in Food Sc. and Nutr. 40: 371-398.

Chapter

5

Beneficial Health Effect of Black Tea: Increasing Evidence

Yogeshwer Shukla[1*], *Sanjoy Kumar Pal*[2], *Annu Arora*[1], *Neetu Kalra*[1] *and Yogendra Kumar Gupta*[1]
[1]Environmental Carcinogenesis Division
Industrial Toxicology Research Centre
P.O. Box No. 80, M.G. Marg
Lucknow-226 001, UP India
e-mail: Yogeshwer_Shukla@hotmail.com
[2]Department of Gastroenterology
Sanjay Gandhi Postgraduate Institute of Medical Sciences
Raibareilly Road
Lucknow - 226 014, UP, India

INTRODUCTION

Tea, an extract of the leaves of the plant *Camellia sinensis*, has been considered a medicine and healthful beverages for ages (Yang et al., 2001). This beverage is consumed worldwide because of its unique aroma, low cost and wide availability (Yeh et al., 2003). The scientific research facilitates that the beneficial effects of tea are due to its polyphenolic components that may affect carcinogen metabolism, free radical scavenging, and formation of DNA adducts (Zhang et al., 2002). The tea plant originated in Southeast Asia and is presently cultivated in about 30 countries around the globe. Although no firm data is available, it appears that tea is universally accepted with a per capita worldwide consumption of approximately 120 ml per day. Of the approximately 2.5 million metric tons of dried tea

*To whom all correspondence should be addressed

manufactured annually, only 20% is green tea, less than 2% is oolong tea and rest is black (regular) tea (Graham, 1992). Green tea is mainly consumed in China, Japan and India and a few countries in North America, Africa and Middle East.

COMPOSITION AND CHEMISTRY OF TEA

Normally the composition of the tea leaf varies with climatic conditions, season, horticultural practices, variety of plant and age of the leaf, i.e., the position of the leaf on the harvested shoot (top two leaves or top three to four leaves). Three main commercial tea beverage varieties are available, and their composition varies according to the manufacturing process involved. In the process of green tea manufacture, the steps include plucking, rapid enzyme inactivation by steaming or pan firing, rolling, and air-drying at high temperature. During the final drying step many new aromatic compounds are formed which provide the characteristic flavour to tea. Withering lowers the moisture level and renders the leaf more amenable to maceration. In this step cell disruption takes place and polyphenol oxidase catalyses the oxidation process. The oxidation is usually allowed to proceed for 45 to 90 minutes, during which some of the catechins are converted to theaflavins and thearubigins which give the characteristic color and taste of black tea (Taneja, 2002).

With regard to the major components, green tea composition is similar to that of the fresh leaf. In the process of black tea production, the steps involved are further oxidizing phases of green tea. More than 500 chemical constituents are reported to be present in tea (Graham, 1992). These include flavanols, flavandiols, flavonoids, phenolic acids, carbohydrates, minerals, proteins, etc. Most of the polyphenols present in green and black teas are flavonols and tannins commonly called as catechins. Some of the major catechins present in tea are: (–)-epigallocatechin gallate (EGCG), (–) -epicatechin-3-gallate (ECG), (–)-epicatechin (EC) (–)-epigallocatechin (EGC), theaflavins (TF) and thearubigins (TR). TFs are astringent compounds, which contribute to the color and taste of the black tea beverage. During black tea manufacture some of the catechins mass is converted in a less well-defined group of compounds known as thearubigins. The TR fraction is a mixed substance of molecular weight 1,000-40,000 and account for 8-20% of total dry matter. Black tea also contains significant aromatic components, which add to its flavor. Presence of vitamin E (24-80 mg/100 gm tea) and vitamin K (300-500 μg/gm) is also reported (Graham, 1992).

PHARMACOLOGICAL PROPERTIES OF TEA

The medicinal use of tea was known long before it was being used as a beverage (Dufresne & Farnworth, 2001). Tea is one of the most widely commonly consumed non-alcoholic beverages and is popular due to its refreshing properties (Weisburger, 1999; Liao et al., 2001). Regular ingestion of tea maintains homeostasis in humans (Hodgson et al., 2001). The pharmacological beneficial effects of tea and its components on health, its role in varied pathophysiological conditions, germicidal, antioxidant, antiatherosclerotic, antimutagenic, anticarcinogenic and detoxifying functions are described briefly.

ANTIOXIDATIVE EFFECTS OF TEA

Tea catechins and its metabolites act as bioantioxidants (Katiyar & Mukhtar, 1997b; Tomita et al., 1997). Tea polyphenols are strong scavengers against superoxide, hydrogen peroxide, hydroxy radicals and nitric oxide produced by various chemicals (Katiyar & Mukhtar, 1997b). Black tea has been identified to act as a powerful chemopreventor of reactive oxygen and nitrogen species (Sarkar & Bhaduri, 2001). Black tea exhibits protective effects against oxidative damage in human red blood cells (Halder & Bhaduri, 1998). Treatment of EGCG to human skin inhibited ultraviolet radiation induced oxidative stress (Katiyar et al., 2001). EGCG also exhibited protective effect against oxidative damage to cellular DNA (Johnson & Loo, 2000). Black and green tea inhibits lipid peroxidative damage in rat liver and kidney (Sano et al., 1995). Besides this, green and black tea ingestion reflected a significant increase of human plasma antioxidant capacity (Serafini et al., 1996). Green tea exhibited antioxidant activity against 22'Azobis (2-amidino propane) dihydrochloride (Yokozawa et al., 2000). Green tea, black tea and EGCG were shown to block the production of oxygen-free radicals derived from IQ in the presence of NADPH-cytochrome P450 reductase (Hasaniya et al., 1997). The scavenging effects of the peroxyl and superoxide radical damage caused by various tea polyphenols is in the following order: EGCG>EGC>ECG>gallic acid>EC>C>caffeine = TF-2>TF-1>TF. Catechin polyphenols could also decrease the peroxynitrite-induced nitration of tyrosine and protect apolipoprotein B-100 of LDL from peroxynitrite-induced modification of critical amino acids, which contribute to its surface charge (Pannala et al., 1997). Black tea and green tea offered protection against oxidative damage to red blood cells caused by a variety of inducers, e.g., H_2O_2, primaquine, 2,2'-azo-bis(2-amidinopropane) dihydrochloride (AAPH), phenylhydrazine (PHx), Cu^{2+}-ascorbic acid, and the xanthine/xanthine oxidase system (Grinberg et al., 1997; Zhang et al., 1997). Oral administration of green or black tea leaf powder inhibited the lipid peroxidation of liver induced by tert-butyl hydroperoxide in rats (Sano et al., 1995). In addition, several reports indicated that tea intake might enhance the levels of antioxidant defense enzymes, such as SOD and catalase. Supplementation of tea polyphenols (0.2%, w/v) to SKH-1 hairless mice for 30 days resulted in increased activities of catalase and glutathione peroxidase in the hepatic, pulmonary, and small bowel (Khan et al., 1992). Administration of aqueous extracts of tea as the sole drinking fluid for 2-4 weeks could increase the UDP-glucuronosyl transferase activity (Bu-Abbas et al., 1995). Previous studies found that after intraperitoneal injections of tea polyphenols in rats, the antioxidant and phase II enzyme activities were elevated (Lee et al., 1995). Theaflavins and its gallate esters exhibit antioxidant properties through action of chelating iron and copper (Miller et al., 1996). Oral feeding of green tea leaves to rats resulted in enhanced SOD activity in serum and catalase activity in liver, and an increased concentration of glutathione in the liver (Lin et al., 1998).

MODULATION OF METABOLIZING/DETOXIFIYING ENZYMES

Tea consumption modulates hepatic drug metabolizing enzymes (Maliakal et al., 2001). Theaflavins prevent DNA damage by inhibiting oxidative stress and suppressing Cytochrome P450 1A1 in cell cultures (Nakamura et al., 2002). Green tea polyphenols have been found to inhibit rat liver microsomal mono-oxygenase activities, including aryl hydrocarbon hydroxylase, 7-ethoxyresorufin O-deethylase and 7-ethoxycoumarin O-deethylase (Wang et al., 1988). Black tea has been found to reduce lipid peroxidative damage and increase glutathione levels in female rats thus enhancing detoxification of toxicants (Sur-Altiner & Yenice, 2000). Another report indicated that treatment of rats with green and black tea for 4 or 6 weeks caused significant induction of cytochrome P450 enzymes, such as CYP1A2, CYP1A1, CYP2B and CYP4A1 (Sohn et al., 1994; Bu-Abbas et al., 1994). Chronic oral administration in mice of 0.2% green tea polyphenols in drinking water for 4 weeks significantly enhanced the activities of phase II detoxifying enzymes, such as glutathione-S-transferase (GST), glutathione peroxidase, catalase and NADPH-quinone oxidoreductase, in the small bowel, lung and liver (Khan et al., 1992).

MODULATION OF IMMUNE FUNCTION

Tea drinking has been associated with increased cell-mediated immune function (Zvetkova et al., 2001). Black tea catechin suppresses the expression of high affinity IgE receptor FC-E RI in human basophillic KU 812 cells (Fujimura et al., 2001). Increase in both number and activity of lymphocytes including NK cells has been reported (Lin et al., 1999). Tea also possesses the property to stimulate the formation of interleukin 2, which has important immuno-regulatory roles (Zvetkova et al., 2001). The phenolic components EGCG and EGC of tea have also been found to regulate the production of pro- and anti-inflammatory cytokines by human leucocytes in vitro (Crouvezier et al., 2001).

ANTIMUTAGENIC ACTIVITY

Microbial systems: Antimutagenic activity of tea polyphenols in microbial systems has been extensively investigated. EGCG and ECG showed inhibitory effects against the mutagenicity of MNNG in *Salmonella typhimurium* TA98 and TA100 with and without rat liver S9 mix (Okuda et al., 1984). EGCG also had a strong inhibitory effect against the mutagenicity of BaP diol expoxide in TA 100 strain without S9 mix. Theaflavins, gallate esters and catechins inhibited mutagenicity of PhIP in *Salmonella typhimurium* TA 98 (Apostolides et al., 1997). The gallate esters of the catechins EC, EGC and EGCG, theaflavonoids TF, TFMG and TFDG and glucose (tannic acid) had low IC_{50} in the 80-250 μM range against mutagenicity of 10 μM PhIP. Non-polyphenolic fraction of green tea suppressed 3-amino-1,4-dimethyl-5H-pyrido[4,3-b]indole (Trp-P-1) or mitomycin C (MMC) induced umu C gene expression in *Salmonella typhimurium* TA1535/psk 1002 in the presence or absence of S9 metabolizing enzyme mixture. (Okai & Higashi-Okai, 1997). Standard black and green tea extracts have been known to inhibit mutagenicity caused by PhIP, in the *Salmonella typhimurium* reverse mutation assay with TA98 containing S9 fraction from the liver of rats induced with alpha-naphthoflavone and Phenobarbital (Apostolides & Weisburger, 1995). The inhibition of the mutagenicity of PhIP by gallate catechins and theaflavins has been found due to restriction of the conversion of the procarcinogen. PhIP, to the proximate carcinogen N-OH-PhIP (Hayatsu et al., 1992). Antimutagenic activity of green and black tea extracts was also observed towards food mutagen MeIQx in the direct plate assay with *Salmonella* *typhimurium* in a *in vitro* gastrointestinal model (Krul et al., 2001). A homogenate of Japanese green tea gave high bioantimutagenic activity against spontaneous mutations resulting from altered DNA-polymerase III in strain NIG1125 of *Bacillus subtilis* (*met his mut-1*) (Kada et al., 1985). EGCG markedly suppressed the direct-acting mutagenicity of MNNG, MNU, captan, and folpet (Hour *et al.*, 1999). Both green tea and black tea decreased the mutagenic activity of MNNG in *Escherchia coli* WP2 (Jain et al., 1989). Pre-incubation together of the MNNG and tea extracts, before the exposure of the cells, also reduced the mutagenic activity of MNNG. In addition tea also exhibited antimutagenic potential against mutagenicity of heterocyclic amines in *Salmonella typhimurium* (Stravic et al., 1996). The antigenotoxic effects of specific tea polyphenols, polyphenon 60 and polyphenon 100 from green tea and polyphenon B from black tea were tested against a battery of genotoxic carcinogens in *Salmonella typhimurium* TA98, TA100 and TA1535 the tester strains (Weisburger et al., 1996; Hara, 1994). The results indicated that tea polyphenols sharply decreased the mutagenicity of a number of aryl-and heterocyclic amines of AFB_{1}_ BaP, DBE and more selectively, of 2-nitropropane, when an induced rat liver S9 fraction was included. The antimutagenicity of catechins was identified against UV-induced in *Escherichia coli* B/r WP2 (Shimoi et al., 1986). Tea extracts inhibited mutagenicity of 1-methyl -1,2,3,4-tetrahydrodiol-b-carboline-3-carboxylic acid on treatment with nitrite in the presence of ethanol (Higashimoto et al., 2000).

Antimutagenic and antioxidative effects of theaflavins from black tea were reported using *Salmonella typhimurium* TA104 (Shiraki et al., 1994). All theaflavins tested, TF, TFA, TFB and TFDG, suppressed mutagenicity induced by H_2O_2, and inhibited peroxidation of rabbit erythrocyte ghosts induced by t-butyl hydroperoxide. EGCG reduced spontaneous mutations in strain NIG1125 of *B. subtilis (met his mut-1)*, due to the inhibition of a function of error-prone DNA replication involving an altered DNA polymerase III (Kada et al., 1985). ECG, EGC and EGCG reduced UVC (254 nm)-induced mutations in *E. coli* B/r WP2 by altering the fidelity of DNA replication (Shimoi et al., 1986). The results also indicated that catechins with a pyrogallol moiety seem to affect the excision-repair system in *E. coli* B/r.

Mammalian cell systems: Crude tea extracts decreased the mutagenic activity of *N-methyl-N'-nitro-N-nitrosoguanidine* in vitro and in intragastric tract of rats (Jain et al., 1989). Black and green tea imparted protection against 2-amino-3-methylimidazo [4,5-f]quinoline-induced DNA adducts and colonic aberrant crypts in the F344 rat (Xu et al., 1996). Anticlastogenic effect of black tea and its two active polyphenols theaflavins and thearubigins was identified towards CP and DMBA in SCE and chromosomal aberration assay *in vivo* in Swiss albino mice (Gupta et al., 2001). Protective effect of green tea has been observed against BaP induced mutations in the liver of big Blue mice (Jiang et al., 2001). Tea and its polyphenols inhibited mitomycin induced micronuclei induction in V79 cells (Liu et al., 1998). Green or black variety of tea, EGCG, and TFG sharply reduced the mutagenicity of IQ and PhIP, and induced DNA repair in rat hepatocytes (Weisburger et al., 1996). ECG and EGCG had inhibitory effects against 6TG-resistant mutations induced by 4NQO in cultured Chinese hamster V79 cells (Kuroda, 1996).

The antimutagenic activity of the catechins was found only when the cells were post-treated with catechins during the mutation expression time after treatment with 4NQO, and were found absent when treated simultaneously with 4NQO and catechins (Kuroda, 1996). Chemopreventive effect of green tea has been attributed against cigarette smoke induced SCE's in humans (Lee et al., 1997). Besides this, in another report, both green tea and black tea retarded 4(methylnitrosamino)-1-(3-pyridyl)-1-butanone bio-activation, DNA me-thylation in A/J mice (Shi et al., 1994). This suggests that the catechins may act intracellularly as bio-antimutagenic blocking agents or suppressive agents. Green tea antioxidant strongly inhibited the increase in SCE and micronuclei induced by fried fish extract and its component MeIQ in V79 or IAR20 cells (Liu, 1990). Tea tannin components inhibited the induction of sister chromatid exchanges and chromosomal aberrations in mutagen-treated cultured mammalian cells (Imanshi et al., 1991). Hot water extracts of green tea effectively suppressed AFB1 induced chromosomal aberrations in bone marrow cells of rats (Ito et al., 1989). Catechins have been found to inhibit tobacco specific 4-(methylnitrosoamino)-1-(3-pyridyl)-1-butanone (NNK) induced DNA strand breaks in rat hepatocytes (Liu & Castonguay, 1991). Green tea suppressed NNK induced 8-OH-guanosine levels, in lung DNA (Xu et al., 1992). Administration of 2% green tea as the sole source of drinking solution for 2 weeks reduced 2-Nitropropane (NP) induced levels of 8-OH-deoxyguanosine adducts in liver nuclear DNA (Hasegawa et al., 1995). Reductions in levels of DNA adducts were also found in rats given extracts of 2% green tea or 1% black tea for 8 weeks before oral administration of IQ (Xu et al., 1996). The amount of IQ and other promutagens decreased in both urine and faeces. Binding of AFB_1 to hepatic nuclear DNA was inhibited in rats given 0.5% instant green tea for 2 or 4 weeks before a single injection of AFB_1 (Quin et al., 1997). The oral administration of 0.2% green tea or 0.1% black tea for 28 days decreased the extent of micronuclei formation in the peripheral blood of mice subsequently treated with BaP and 3-methylcholanthrene (Sasaki et al., 1993). It was also found that green tea given by stomach perfusion had distinct inhibitory effects on micronuclei in the colon crypt cells of C57BL mice (Zao et al., 1992).

ANTICARCINOGENIC ACTIVITY

Apart from other beneficial health effects of tea it is also known to possess anticarcinogenic properties. Substantial information including both experimental and epidemiological studies is available showing the anticarcinogenic effects of both tea extracts and polyphenols (Katiyar & Mukhtar, 1996; Kuroda & Hara, 1999; Isemura et al., 2000). Several mechanisms are put forward to explain the chemopreventive effect of tea. These pathway include: signal transduction pathways leading to activator protein - 1 (AP-1) and/or nuclear factor kappa (NF-kappaB). AP-1 and NF-kappaB are transcription factors which induce, cell transformation and tumor promotion and are influenced differentially by the MAP kinase pathways (Park & Dong, 2003).

Skin tumors: Convincing evidence has been documented showing that topical application or oral feeding of a polyphenolic fraction from tea extract and individual catechin derivatives, had anticarcinogenic effects in animal skin experiments (Wang et al., 1994; Katiyar & Mukhtar, 1997a; Javed et al., 1998). Black tea, decaffeinated black tea, and decaffeinated green tea exhibited inhibitory effects of on ultraviolet B light-induced skin carcinogenesis in DMBA -induced SKH-1 mice (Wang et al., 1994). Caffeine also contributed in an important way to the inhibitory effects of green tea and black tea on UVB-induced complete skin carcinogenesis (Huang et al., 1997). Aqueous extracts of green tea inhibited tumor initiation by UVB (ultraviolet B) light for 10 days and TPA for 25 weeks, and also tumor initiated by DMBA with tumor promotion by TPA or UVB (Wang et al., 1992b). Oral administration of black tea as the drinking fluid for 11 weeks decreased the number of non-malignant tumors (kelatoacanthoma) and squamous cell carcinoma in mice, which had been produced by irradiation with UVB (Lu et al., 1997). Black tea inhibited proliferation of kelatoacanthomas and squamous cell carcinomas and enhanced apoptosis in nonmalignant and malignant skin tumors, and resulted in a reduction in the number and volume of the tumors. Topical application of Black tea polyphenols inhibited DMBA initiated and TPA promoted skin tumorigenesis in Swiss albino mice (Javed et al., 1998).

Tumors in gastrointestinal tract: Black and green tea have been found to inhibit N-nitrosomethylbenzylamine (NMBA) induced oesophageal tumorogenesis in rats (Wang et al., 1995). Inhibitory effects were observed when tea was given during and after treatment, suggesting that tea components can inhibit both initiation and post initiation events in oesophageal tumorigenesis. Theaflavin component was also found to be effective in suppressing NMBA induced forestomach tumor formation (Kuroda & Hara, 1999). Furthermore, the chemopreventive effects of tea were also attributed against human oral precancerous lesions (Li et al., 1999). Administration of tea polyphenols retarded the growth of squamous cell carcinoma *in vitro* (Elattar & Virji, 2000). Tea extracts and catechins, especially EGCG, inhibited the growth of tumors induced by carcinogens in the gastrointestinal tract (Morse et al., 1997; Elattar & Virji, 2000). Green tea infusion, given as the sole source of drinking water to A/J mice, decreased the forestomach tumor multiplicity induced by NDEA given p.o. once a week for 8 weeks (Kinlen et al., 1988). However, pooled analysis of two prospective studies in Japan could find no relevance between green tea consumption and risk of gastric cancer (Koizumi et al., 2003).

Mammary gland tumors: Black tea inhibited mammary gland carcinogenesis induced by DMBA in rats fed with saturated fatty diets (Rogers et al., 1998). The survival rates of rats maintained on a diet containing black tea were higher than those of rats maintained on a normal control diet. The incidence of mammary adenocarcinomas was also found to be decreased. Green tea has been found to inhibit PhIP and DMBA induced mammary tumors (Hirose et al., 1995; Tanaka et al., 1997).

Hepatocarcinoma: A number of studies provide evidence for the chemopreventive efficacy of green and black tea against induction of hepatic tumors. Green tea exhibited significant protection against DEN induced hepatocarcinogenesis (Li, 1991). Studies have shown that epicatechin derivatives of tea reduced DEN induced liver precancerous lesions, variant cell foci, N-ras overexpressed lesion and hepatic nodule formation in rats (Mao, 1993). Furthermore, administration of EC, EGC, ECG, EGCG, black tea extract or oolong tea extract at concentrations of 0.05 or 0.1% in drinking water for 6 weeks reduced DEN induced hepatocarcinoma in F344 rats (Matsumoto et al., 1996). The growth of spontaneous hepatomas in mice and human hepatoma cell line were inhibited by EGCG at concentrations of

0.05% and 0.1% in drinking water (Nishida et al., 1994). Studies of Jia et al. (2002) have shown that tea polyphenols and tea pigments are effective in preventing the precancerous liver lesions in rat and modulation of cell cycle. Recent studies of (Kuo and Lin, 2003) have shown that EGCG inhibited the proliferation of liver cancer cell line, Hep G2 by inducing apoptosis and blocking cell cycle progression in the G1 phase. EGCG significantly increased the expression of p53 and p21/WAF1 protein, and this contributes to cell cycle arrest.

Lung cancer: Green tea catechins and black tea theaflavins inhibit collagenase in mouse lung carcinoma cells (Sazuka et al., 1995). Furthermore black and green tea suppressed spontaneous formation of lung tumors and rhabdomyosarcomas in A/J mice (Landau et al., 1998, Xu et al., 1992). Inhibition of N-nitrosodiethylamine and 4-(methylnitrosoamine)-1-(3-pyridyl)-1-butanone-induced and pulmonary tumorigenesis was observed by administration of tea (Wang et al., 1992a). Anticarcinogenic effect of black tea against Diethylnitrosoamine (DEN) induced pulmonary adenomas and adenocarcinoma has been observed in Swiss albino mice (Shukla & Taneja, 2002). EGCG prevented cisplatin-induced lung tumorigenesis (Mimoto et al., 2000). Similar results were also observed from green or black tea extract, EGCG, or theaflavins which were effective against lung tumorigenesis induced by the tobacco carcinogen 4-(methylnitrosamino)-1-(3-pyridyl)-1-butanone (NNK) (Shi et al., 1994). The progression of adenoma to adenocarcinoma and the cell proliferation rate in adenomas were also suppressed by black tea treatment (Kohlmeirer et al., 1997) Besides this, oral feeding of black tea extract has been found to exhibit protection against lung tumorigenesis induced by NNK in rats, and that caffeine may be an important constituent involved in inhibition of lung carcinogenesis (Chung et al., 1998).

Transplantable tumors: The crude catechins extracted from tea have been reported to inhibit growth of 3-methylcholanthrene (3-MC) induced solid tumors, mouse sarcomas and Ehrlich Ascites (EA) tumors (Hara et al., 1994). Moreover long term feeding of tea flavonoids suppressed the growth of inoculated tumors. Recent study has shown the anticarcinogenic activity of black tea on growth of transplantable EA and solid tumors induced by DMBA in Swiss albino mice (Javed & Shukla, 2000). Black tea administration to Ehrlich ascites carcinoma (EAC)-bearing Swiss albino mice caused a significant decrease in the tumor cell count in a dose dependent manner (Bhattacharyya et al., 2003).

Inhibition of tumor promotion: Tea polyphenols are known to exhibit significant antitumor promoting activity directly or by inhibiting the action of tumor promoters. Theaflavins and EGCG inhibit tumor promoter-induced activation and cell transformation (Dong et al., 1997). EGCG inhibited tumor promotion by teleocidin in multi-step carcinogenesis (Yoshizawa et al., 1987). EGCG inhibited tumor promotion in the gastrointestinal tract using a model system of mouse duodenal carcinogenesis (Katiyar & Mukhtar, 1996). Black tea polyphenols exhibited inhibitory activity against TPA promoted skin carcinogenesis (Javed et al., 1998). Tea has also been shown to block the promotion of DMBA-induced mammary tumorigenesis (Rogers et al., 1998). The suppression of induced ornithine decarboxylase (ODC), cyclooxygenase and lipooxygenase activities has been suggested to be the possible mode of action by which polyphenols exert their antitumor promoting effect (Yang et al., 2000a). The topical application of green tea polyphenol fraction resulted in the inhibition of TPA mediated tumor promotion and ODC activity (Huang et al., 1992). Black tea inhibited tumor growth and prevented metastasis in 3-methylcholanthrene induced solid tumor model in mice (Das et al., 2002). Green tea polyphenols pretreatment also inhibited testosterone-mediated induction of ODC in human prostate cancer cells (Gupta et al., 1999). Tea polyphenols depicted inhibitory action on TPA induced promotion and epidermal interleukin-1a mRNA activity in mice (Katiyar et al., 1995). It has also been postulated that tea polyphenolic ingredients block the interaction of tumor promoters, and growth hormones with their receptors, which may account for its observed antitumor-promoting activity (Katiyar & Mukhtar, 1996; Kuroda & Hara, 1999).

Inhibition of tumor proliferation and mitotic signal transduction: Topical application of green tea polyphenols to mouse skin was found to inhibit TPA-induced ODC activity in a dose-dependent manner (Agarwal et al., 1992). EGCG and TF-3 inhibited TPA-induced transformation, PKC activation and AP-1 binding activities in mouse fibroblast cells (Lee & Lin, 1997; Chen et al., 1999). Inhibitory effects of green tea polyphenols on TPA-induced mouse epidermal lipooxygenase and cyclooxygenase activities *in vivo* have been observed (Katiyar et al., 1992). EGCG and theaflavins inhibited epidermal growth factor/TPA-induced cell transformation in a dose-dependent manner (Dong et al., 1997). The molecular mechanism of the antipromotion activity of EGCG and theaflavins involves blocking of the AP-1-dependent transcriptional activity and DNA binding activity. The inhibition of AP-1 activation occurs through the inhibition of a c-Jun NH_2-terminal kinase-dependent, but not an extracellular signal-regulated protein kinase (ErK) 1-dependent or ErK2-dependent pathway (Nomura et al., 2000; Lin & Liang, 2000). The molecular mechanisms of antitumor promotion might operate by blocking the mitotic signal transduction pathway (Lin & Lee, 1995; Lin et al., 1999a). EGCG and TF-3 strongly inhibited the kinase activity of EGF and PDGF receptors induced by EGF or PDGF, respectively (Liang et al., 1997). These results suggested that EGCG and TF-3 have an antiproliferation effect on tumor cells, and that the molecular mechanisms of this effect may block the growth factor binding to its receptor and subsequently through inhibition of receptor kinase activity, and suppress mitogenic signal transduction. Tea polyphenols might suppress mitogenic signal transduction and other unknown proliferative signals, which might have an inhibitory effect on proliferation and cell cycle progression, even on apoptosis.

Induction of apoptosis: Tea polyphenols and their metabolites induce apoptosis and cell cycle arrest and inhibit the growth of human cancer cell lines (Yang et al., 1998; Zhang et al., 2000). Black tea polyphenols, green tea extract and EGCG have been shown to inhibit the growth of rat hepatoma, mouse erythroleukaemia and the growth of several human cancer cell lines, such as MCF-7 breast carcinoma, HT-29 colon carcinoma, A-427 lung carcinoma and UACC-375 melanoma (Valcic et al., 1996). The induction of apoptosis by catechins has also been demonstrated in human lymphoid leukaemia cells (Hibasami et al., 1996). Recent studies have shown that tea polyphenol-induced apoptosis is mitochondrial-targeted, caspase 3 gene mediated (Hsu et al., 2003). Tea polyphenols regulate c-jun phosphorylation and decrease H_2O_2 production in transformed and non-transformed human bronchial cell lines through mechanisms of cell growth inhibition and apoptosis induction (Yang et al., 2000b). In another study the growth inhibition by EGCG in SV-40 virally transformed human fibroblast cells W138 has been demonstrated through induction of apoptosis (Chen et al., 1998). Tea polyphenols such as EGCG and ECG also inhibit the growth of human lung cancer cell line PC-9 (Okabe et al., 1997). The study further demonstrated that growth inhibition was accompanied by G2/M phase arrest of cell cycle. Moreover, the inducing effect of EGCG on apoptosis has been enhanced by several cancer preventive compounds such as sulinidac and tamoxifen (Suganama et al., 1999). The exposure of human cancer cells KATO III to black tea extracts and TFs resulted in inhibition and induction of apoptosis with characteristic fragmentation (Hibasami et al., 1998). The growth inhibition by various tea polyphenols has been demonstrated in human cancer cells H-161 and H-1299 (Yang et al., 1998). Green tea and its constituents selectively induce apoptosis in oral carcinoma cells, while EGCG was able to inhibit their growth and invasion (Hsu et al., 2002). However, how normal epithelial cells survive in the presence of polyphenols is unknown. Recent studies have suggested that p57/KIP2 is a determinant pro-survival factor for cell protection from tea polyphenol-induced apoptosis (Hsu et al., 2002).

Anti-inflammatory effects: Black tea polyphenols application resulted in significant inhibition of TPA-caused inflammatory responses and suppression of epidermal ODC and of COX enzyme activities (Katiyar & Mukhtar, 1997b). Nitric oxide (NO) radical has a wide biological role in modulating physiological and patho-physiological processes (Moncada et al., 1992). During infection and inflammation, the formation of NO is increased, which might promote neoplastic conversion. Tea polyphenols regulate cyclooxygenase- and lipoxygenase-dependent metabolism of arachidonic acid in human colon mucosa and colon tumor tissues (Hong et al., 2001). Both EGCG and TF-3 inhibited the induction of NO in thioglycollate-elicited and lipopoly-saccharide (LPS)-activated peritoneal or RAW2647 macrophages (Lin & Lin, 1997; Chan et al., 1997; Lin et al., 1999b). EGCG decreased the activity and protein levels of inducible nitric oxide synthase (iNOS) by reducing the expression of iNOS mRNA and that the reduction could occur as a result of prevention of binding of the nuclear factor-kB to the iNOS promoter, thereby inhibiting the induction of iNOS transcription (Lin & Lin, 1997).

PREVENTION OF CORONARY HEART DISEASE

Black tea consumption has been found to reverse endothelial dysfunction in patients with coronary artery disease (Duffy et al., 2001). The anti-hypertensive effect of tea has been attributed to different components of tea. Crude catechin (0.5%) supplemented in the diet was shown to reduce blood pressure in rat (Vinson, 2000). Tea exhibits proconvulsive effect in mice (Gomes et al., 1999). Drinking of green and black tea has been associated with decrease in the incidence of high blood pressure incidence (Hodgson et al., 1999). Tea tannin (2-4 mg orally) could also reduce the systolic blood pressure in rats. Investigation proved that tea components, especially TFDG and EGCG, showed beneficial effect (Hertog et al., 1997). Tea polyphenols and caffeine may also ameliorate the development of hypertension by improving the renal circulatory state (Duffy et al., 2001) as a result of its direct action on the kidney, including the activation of the kinin-kallikrein-prostaglandin system in the kidney. Tea drinking depressed corpulence by increasing metabolic rate and degradation of fat. Black green and oolong tea extracts modulate lipid metabolism in hyperlipidemia rats (Hertog et al., 1997). Tea could reduce the cholesterol absorption from the intestine by reducing solubility of cholesterol in mixed micelles (Duffy et al., 2001). The effect of black tea on soluble P-selectin provides a potential mechanism for cardiovascular benefits of regular ingestion of tea (Hodgson et al., 2001). Among the various catechins, the gallate ester type catechins (EGCG, ECG) were most effective in precipitating micellar cholesterol than free catechins EC and EGC (Hertog et al., 1997). Recent epidemiological studies have shown potential protective effect of black tea consumption in relation to coronary heart disease (CHD) in Saudi Arabia (Hakim et al., 2003).

GERMICIDAL AND ANTIVIRAL ACTIVITY

Investigations have proved that tea exhibits germicidal and antiviral acitivity (Toda et al., 1989). Crude catechins, gallated catechins and TFs, TFDG, TF monogallate A and B inhibited the growth of both spores and vegetative cells of *C. botulinum.s.* ECG and EGCG possessed the inhibitory activity against hemolysin excreted by *Vibro cholera,* the toxin excreted by *Staphylococcus aureus* bacteria and damage to CHO cells induced by *Cholera toxin* (Toda et al., 1990). Tea polyphenols have been found to retard the infectivity of influenza virus (Nakayama et al., 1993). Besides this EGCG has been found to exhibit inhibitory effects on the life cycle of human immunodeficiency virus type 1 [HIV-1] (Yamaguchi et al., 2002). Black tea extract, thearubigin fraction, counteracts and protects against the adverse effect of tetanus toxin in mice (Satoh et al., 2001). Recent studies indicate that epicatechin gallate in tea may be a valuable therapeutic agent against *H. pylori* infection (Yanagawa et al., 2003).

REGULATION OF INTESTINAL MICROFLORA

Extracts of Black tea, Japanese green tea and Chinese tea are known to inhibit the growth of various bacteria causing diarrhoeal diseases (Shetty et al., 1994). Tea polyphenols exhibit protection against intestinal disorders as shown by the influence of tea on propagation of *Bifidobacterium* and *Clostridium* (Okubo et al., 1991). All tea samples tested showed protective activity against diarrohoea causing bacteria including *Staphylococcus aureus, S. epidermidis, Vibrio cholerae O1, V. cholerae non O1. V. parahaemolyticus,*

V. mimicus, *Campylobacter jejuni* and *Plesiomonas shigelloides* (Toda et al., 1989).

PREVENTION OF DENTAL CARIES

Tea drinking has been associated with decreased incidence of dental caries (Rashed & Haider, 1998; Hamilton-Miller, 2001). Experimental studies on rats revealed that black tea administration could prevent and reduce caries lesions by inhibition of adherence of the pathogenic bacteria on teeth surface and thus exhibiting antiplaque activity (Touyz & Amsel, 2001). Investigations had showed that TF, theaflavin digallate ECG, GCG and EGCG strongly inhibited water-soluble and-insoluble glucan synthesis by glucotransferase enzyme excreted by plaque causing bacteria (Hattori et al., 1990).

CONCLUSION

Botanicals have been used for the treatment of various human diseases throughout the history (Park & Pezzuito, 2002). In addition, botanicals play a role in disease prevention. Tea is an important dietary source of flavanols and flavonols (McKay & Blumberg, 2002). *In vitro* and animal studies provide strong evidence that tea polyphenols may possess the bioactivity to affect the pathogenesis of several chronic diseases, especially cardiovascular disease and cancer. The bioavailability and biotransformation of tea polyphenols, however, are key factors limiting these activities *in vivo* (Ioannides & Yoxall, 2003). The inhibition of tumorigenesis by black or green tea preparation has been demonstrated in animal models on different organ sites such as skin, lung, oral cavity, esophagus, forestomach, stomach, small intestine, colon, pancreas, and mammary gland. Epidemiological studies, however, have not yielded clear conclusion concerning the protective effect of tea consumption in humans. The discrepancy between the results from humans and animal models could be due to, 1) much higher doses of tea used in animals in comparison to human consumption, 2) differences in causative factors between the cancers in humans and animals, and 3) confounding factors limiting the power of epidemiological studies to detect an effect (Yang et al., 2002). Other factors that could also play a role are socioeconomic and lifestyle factors as well as the inadequate methodology to define tea preparation and intake. It is also possible that tea may be effective against only specific types of cancer caused by certain etiological factors. Common use of antimutagens and anticarcinogens in everyday life is an effective measure for prevention of cancer and gastric diseases. Antioxidant properties of tea have vast potential as a protective agent against diverse toxic effects, which improve the antioxidation status and thereby reduce the risk of many diseases.

REFERENCES

Agarwal, R., Katiyar, S.K., Zaidi, S.I.A. and Mukhtar, H. (1992). Inhibition of tumor promoter caused induction of ornithine decarboxylase activity in SENCAR mice by polyphenolic fraction isolated from green tea and its individual epicatechin derivatives. Cancer Res., 52: 3582-3588.

Apostolides, Z., Balentine, D.A., Harbowy, M.E., Hara, Y. and Weisburger, J.H. (1997). Inhibition of PhIP mutagenicity by catechins and by theaflavins and gallate esters. Mutat. Res., 389: 167-172.

Apostolides, Z., and Weisburger, J.H. (1995). Screening of tea clones for inhibition of PhIP mutagenicity. Mutat. Res., 6: 219-225.

Bhattacharyya, A., Choudhuri, T., Pal, S., Chattopadhyay, S.K., Dutta, G., Sa G. Das, T. (2003) Apoptogenic effect of black tea on Ehrich's ascites carcinoma cell. Carcinogenesis, 24: 75-80.

Bu-Abbas, A., Clifford, M.N., Ioannnides, C., and Walker, R. (1995). Stimulation of rat UDP-glucuronosyl transferase activity following treatment with green tea. Food. Chem. Toxicol., 33: 27-30.

Bu-Abbas, A., Clifford, M.N., Walker, R. and Ioannides, C. (1994). Selective induction of rat CYP1 and CYP4 protein and peroxisomal proliferation by green tea. Carcinogenesis 15: 2575-2579.

Chan, M.M., Fong, D., Ho, C.T. and Huang, H.I. (1997). Inhibition of inducible nitric oxide synthetase gene expression and enzyme expression by epigallocatechin gallate a natural product from green tea. Biochem. Pharmacol., 54: 1281-1286.

Chen, Z.P., Schell, J.B., Ho, C.T. and Chen, K.Y. (1998). Green tea epigallocatechin gallate shows a pronounced growth inhibitory effect on cancerous cells but not on their normal counterparts. Cancer Lett., 129: 173-179.

Chen Y.C., Liang, Y.C., Lin-Shiau, S.Y., Ho, C.T. and Lin, J.K. (1999). Inhibition of TPA-induced PKC and AP-1 binding activities by theaflavin-3,3'-digallate from black tea in NIH3T3 cells. J. Agric. Food Chem., 47: 1416-1421.

Chung, F.L., Wang, M., Rivenson, A., latropoulos, M.J., Reinhardt, J.C., Pittman, B., Ho, C.T. and Amin, S.G. (1998). Inhibition of lung carcinogensis by black tea in Fischer rats treated with a tobacco-specific carcinogen: caffeine as an important constituent. Cancer Res., 58: 4096-4101.

Crouvezier, S., Powell, B., Keir, D. and Yaqoob, P. (2001). The effects of phenolic components of tea on production of pro- and anti-inflammatory cytokines by human leukocytes in vitro. Cytokine, 13: 280-286.

Das, M., Sur, P., Gomes, A., Vedasiromoni, J.R. and Ganguly, D.K. (2002). Inhibition of tumor growth and inflammation by consumption of tea. Phytother Res., 16 (Suppl 1): S40-S44.

Dong, Z., Ma, W.Y., Huang, C. and Yang, C.S. (1997). Inhibition of tumor-promotor induced activator protein 1 activation and cell transformation by tea polyphenols, (–) -epigallocatechin gallate, and theaflavins. Cancer Res., 57: 4414-4419.

Duffy, S.J., Keaney, J.F Jr., Holbork., M., Gokce, N., Swerdloff, P.L., Frei, B. and Vita, J.A. (2001). Short- and long-term black tea consumption reverses endothelial dysfunction in patients with coronary artery disease. Circulation, 104: 151-156.

Dufresne, C.J. and Farnworth, E.R. (2001). A review of latest research findings on the health promotion properties of tea. J. Nutr. Biochem., 12: 404-421.

Elattar, T.M. and Virji, A.S. (2000). Effect of tea polyphenols on the growth of squamous cell carcinoma cells in vitro. Anticancer Res., 20: 3459-3465.

Fujimura, Y., Tachibana, H. and Yamada, K. (2001). A tea catechin suppressess the expression of high affinity IgE receptor FC—E RI in human basophillic KU 812 cells. J. Agric. Food. Chem. 49: 2257-2531.

Gomes, A., Das, M., Vedasiromoni, J.R. and Ganguly, D.K. (1999). Proconvulsive effect of tea (*Camellia sinensis*) in mice. Phytother. Res., 13: 376-379.

Graham, H.N. (1992). Green tea composition consumption and polyphenoi chemistry. Prev. Med., 21: 334-350.

Grinberg, L.N., Newmark, H., Kitrossky, N., Rahamim, E., Chevion, M. and Rachmilewitz, E.A. (1997). Protective effects of tea polyphenols against oxidative damage to red blood cells. Biochem. Pharmacol., 54: 973-978.

Gupta, S., Chaudhuri, T., Ganguly, D.K. and Giri, A.K. (2001). Anticlastogenic effects of black tea (World blend) and its two active polyphenols theaflavins and thearubigins in vivo in Swiss albino mice. Life Sci., 69: 2735-2744.

Gupta, S., Ahmad, N., Mohan, R.R., Hussain, M.M. and Mukhtar, H. (1999). Prostate cancer chemoprevention by green tea: *in vitro* and in vivo inhibition of testosterone-mediated ornithine decarboxylase. Cancer Res., 59: 2115-2120.

Hakim, I.A., Alsaif, M.A., Alduwaihy, M., Al-Rubeaan, K., Al-Nuaim. A.R., and Al-Attas, O.S. (2003). Tea consumption and the prevalence of coronary heart disease in Saudi adults: results from a Saudi national study. Prev. Med., 36: 64-70.

Halder, J. and Bhaduri, A.N. (1998). Protective effect of black tea against oxidative damage in human red blood cells. Biochem. Biophys. Res. Commun., 244: 903-907.

Hamilton-Miller, J.M. (2001) Anti-cariogenic properties of tea (*Camellia sinensis*). J. Med. Microbiol. 50: 299-302. Review.

Hara, Y., Ho, C.T., Osawa, T., Huang, M.T. and Rosen, T.R. (Eds). Prophylatctic functions of tea polyphenols In: Food Phytochemicals for Cancer Prevention II. Teas. Spieces and Herbs. ACS Symp. Series 547: Washington, DC, pp. 34-50.

Hasaniya, N., Youn, K., Xu, M., Hernaez, J. and Dashwood, R. (1997). Inhibitory activity of green and black tea in a free radical-generating system using 2-amino-3-methylimidazo[4,5-f] quinoline a substrate. Jpn. J. Cancer. Res., 88: 553-558.

Hasegawa, R., Chujo, T., Sai-Kato, K., Umemura, T., Tanimura, A. and Kurokawa, Y. (1995). Preventive effects of green tea against liver oxidative DNA damage and hepatotoxicity in rats treated with 2-nitropropane. Food. Chem. Toxicol., 33: 961-970.

Hattori M., Kusumoto I.T., Namba T., Ishigami T. and Hara Y. (1990). Studies on dental caries prevention by traditional medicines. Part XIII. Effect of tea polyphenols on gluco synthesis by glucosyltransferase from *Streptococcus mutans*. Chem. Pharm. Bull. 38: 717-720

Hayatsu, H., Inaba, N., Kakutani, T., Arimoto, S., Negishi, T., Mori, K., Okuda, T. and Sakata, I. (1992). Suppression of genotoxicity of carcinogenesis by (–)-epigallocatecin gallate. Prev. Med. 21: 370-376.

Hertog, M.G., Sweetnam, P.M., Fehily, A.M., Flwood, P.C. and Kromhout, D. (1997). Antioxidant flavonols and ischemic heart disease in a Welsh population of men: the caerphilly study. Am. J. Clin. Nutr., 65: 1489-1494.

Hibasami, H., Achiwa, Y., Fujikawa, T. and Komiya T. (1996). Induction of programmed cell death (apoptosis) in human lymphoid leukemia cells by catechin compounds. Anticancer Res., 16: 1943-1946.

Hibasami, H., Komiya, T., Achiwa, Y., Ohnishi, K., Kojima, T., Nakanishi, K., Sugimoto, Y., Hasegawa, M., Akatsuka, R. and Hara, Y. (1998). Black tea theaflavins induce programmed cell death in cultured human cancer cells. Int. J. Mol. Med., 1: 725-727.

Higashimoto, M., Akada, Y., Sato, M., Kimouchi, T. and Kuwahara, Y. (2000) Inhibitory effects of tea extracts on mutagenecity of 1-methyl-1,2,3,4-tetrahhydrodiol-b-carboline-3-carboxylic acid on treatment with nitrite in the presence of ethanol. Food. Chem. Toxicol., 38: 7-13.

Hirose, M., Akagi, K., Hasegawa, R., Yaono, R.M., Satoh, T., Hara, Y., Wakabayashi, K. and Ito, N. (1995). Chemoprevention of of 2-amino-1-methyl-6-phenylimidazo [4,5-b]-pyridine induced mammary gland carcinogenesis by antioxidants in F344 rats. Carcinogenesis, 16: 217-221.

Hodgson, J.M., Puddey, I.B., Burke, V., Beilin, L.J. and Jordan, N. (1999). Effects on blood pressure of drinking green and black tea. J. Hypertens., 17: 457-463.

Hodgson, J.M., Puddey, I.B., Mori, T.A., Burke, V., Baker, R.I. and Beilin, L.J. (2001). Effects of regular ingestion of black tea on haemostasis and cell adhesion molecules in humans. Eur. J. Clin. Nutr., 55: 881-886.

Hong, J., Smith, T.J., Ho, C.T., August, D.A. and Yang, C.S. (2001). Effects of purified green and black tea polyphenols on cyclooxygenase and lipoxygenase-dependent metabolism of arachidonic acid in human colon mucosa and colon tumor tissues. Biochem Pharmacol., 62: 1175-1183.

Hour, T-C., Liang, Y-C., Chu, I-S and Lin, J-K. (1999). Inhibition of eleven mutagens by various tea extracts, (–) epigallocatechin-3-gallate, gallic acid and caffeine. Food. Chem. Toxicol., 37: 569-579.

Hse, S.D., Singh, B.B., Lewis, J.B., Broke, J.L., Dickinson, D.P., Drake, L., Caughman, G.B., and Schuster, G.S. (2003). Chemoprevention of oral cancer by green tea. Gen Dent. 50: 140 -146.

Hsu, S., Lewis, J., Singh, B., Schoenlein, P., Osaki, T., Athar, M., Porter, A.G., and Schuster, G. (2003). Green tea polyphenol targets the mitochondria in tumor cells induced caspase 3-dependent apoptosis. Anticancer Res., 23: 1533 -1539.

Hsu, S., Yu, F.S., Lewis, J., Singh, B., Borke, J., Osaki, T., Athar, M., and Schuster, G. (2002). Induction of p57 is required for cell survival when exposed to green tea polyphenol. Anticancer Res., 22: 4115-4120.

Huang, M.T., Ho, C.T., Wang, Z.Y., Ferrare, T., Finnegan Olive, T., Lou, Y.R., Mitchell, J.M., Laskin, J.D., Newmark, H. and Yang, C.S. (1992). Inhibitory effect of topical application of a green tea polyphenol fraction on tumor initiation and promotion in mouse cell. Carcinogenesis, 13: 947-954.

Huang, M.T., Xie, J.G., Wamg, Z.Y., Ho, C.T., Lou, Y.R., Wang, C.X., Hard, G.C. and Conney, A.H. (1997). Effects of tea, decaffeinated tea, and caffeine on UVB light-induced complete carcinogenesis in SKH-1 mice: demonstration of caffeine as a biologically important constituent of tea. Cancer Res., 57: 2623-2629.

Imanshi, H., Sasaki, Y.F., Ohta, T., Watanabe, M., Kato, T. and Shirazo, Y. (1991). Tea tannin components modify the induction of sister chromatid exchanges and chromosome aberrations in mutagen-treated cultured mammalian cells and mice. Mutat. Res., 259: 79-87.

Ioannides, C. and Yoxall, V. (2003) Antimutagenic activity of tea: role of polyphenols. Curr. Opin. Clin. Nutr. Metab. Care., 6: 649-656.

Isemura, M., Saeki, K., Kimura, T., Hayakawa, S., Minami, T. and Sazuka, M. (2000). Tea catechins and related polyphenols as anticarcinogens. Biofactors., 13: 81-85.

Ito, Y. Ohnishi, S. and Fujie, K. (1989). Chromosome aberrations induced by aflatoxin B1 in rat bone marrow cells in vivo and their suppression by green tea. Mutat. Res., 222: 253-261.

Jain, A.K., Shimoi, K., Nakamura, Y., Kada, T., Hara, Y. and Tomita, I. (1989). Crude tea extracts decrease the mutagenic activity of N-methyl-N'-nitro-N-nitrosoguanidine in vitro and in intragastric tract of rats. Mutat. Res. 210: 1-8.

Javed, S., Mehrotra, K. and Shukla, Y. (1998). Chemopreventive effects of black tea polyphenols in mouse skin model of carcinogenesis. Biomed. Env. Sci., 11: 307-313.

Javed, S. and Shukla, Y. (2000) . Effect of black tea extract on transplantable and solid tumors in Swiss albino mice. Biomed. Env. Sci. 10: 213-218.

Jia, X., Han, C. and Chen, J. (2002) Effect of tea on preneoplastic lesion and cell cycle regulators in rat liver. Cancer Epidemiol. Biomarkers Prev., 12: 1663-1667.

Jiang, T., Glickman, B.W. and deBoer, J.G. (2001). Protective effect of green tea against benzo (a) pyrene-induced mutations in the liver of Big Blue mice. Mutat. Res., 480: 147-151.

Johnson, M.K. and Loo, G. (2000). Effects of epigallocatechin gallate and quercetin on oxidative damage to cellular DNA. Mutat. Res., 459: 211-218.

Kada, T., Kaneko, K., Matsuzaki, S., Matsuzaki, T. and Hara Y. (1985). Detection and chemical identification of natural bioantimutagens. A case of the green tea factor. Mutat. Res., 150: 127-132.

Katiyar, S.K. and Mukhtar, H. (1996). Tea in chemoprevention of cancer: Epidemiologic and experimental studies (Review). Int. J. Oncol. 8: 221-238.

Katiyar, S.K. and Mukhtar, H. (1997a). Tea antioxidants in cancer chemoprevention. J. Cell. Biochem., 27: 59-67.

Katiyar, S.K. and Mukhtar, H. (1997b). Inhibition of phorbol ester tumor promoter in 12-O-tetradeconylphorbol-13-acetate-caused inflammatory responses in SENCAR mouse skin by black tea polyphenols. Carcinogenesis, 18: 1911-1916.

Katiyar, S.K., Afaq, F., Perez, A. and Mukhtar, H. (2001). Green tea polyphenol (–)-epigallocatechin-3-gallate treatment of human skin inhibits ultraviolet radiation induced oxidative stress. Carcinogenesis, 22: 287-294.

Katiyar, S.K., Agarwal, R., Wood, G.S., and Mukhtar, H. (1992). Inhibition of 12-O-tetradeconylphorbol-13-acetate caused tumor promotion in 7,12-dimethylbenz [a]anthracene initiated SENCIAR mouse skin by polyphenolic fraction isolated from green tea. Cancer Res., 52: 6890-6897.

Katiyar, S.K., Rupp, C.O., Korman, N.J., Agarwal, R. and Mukhtar, H. (1995). Inhibition of 12-O-tetradeconylphorbol-13-acetate and other skin tumor promoter-caused induction of epidermal interleukin-1mRNA and protein expression in SENCAR mice by green tea polyphenols. J. Invest Dermatol., 105: 394-398.

Khan, S.G., Katiyar, S.K., Agarwal, R. and Mukhtar, H. (1992). Enhancement of antioxidant and phase II enzyme by oral feeding of green tea polyphenols in drinking water to SKH-1 hairless mice: possible role in cancer chemoprevention. Cancer Res., 52: 4050-4052.

Kinlen L.J., Willows A.N., Guldblatt P. and Yudkin J. (1988). Tea consumption and cancer. Br. J. Cancer 58: 397-401.

Kohlmeirer, L., Weterings, K.G., Steck, S. and Kok, F.J. (1997). Tea and cancer prevention: an evaluation of epidemiological literature. Nutr. Cancer, 27: 1-13.

Koizumi, Y., Tsubono, Y., Nakaya, N., Nishino, Y., Shibuya, D., Matsuoka, H., and Tsuji, I. (2003). No association between green tea and the risk of gastric cancer: pooled analysis of two prospective studies in Japan. Cancer Epidemiol. Biomarkers Prev., 12: 472-473.

Krul, C., Luiten-Schuite, A., Tenfelde, B., vanOmmen, B., Verhagen, H. and Havenaar, R. (2001). Antimutagenic activity of green and black tea extracts studied in a dynamic in vitro gastrointestinal model. Mutat. Res., 474: 71-85.

Kuo, P.L. and Linn, C.C. (2003). Green tea constitute (–)-epigallocatechin-3-gallate inhibits Hep G2 cell proliferation and induces apoptosis through p53-dependent and Fas-mediated pathways. J. Biomed. Sci., 10: 219-227.

Kuroda, Y. and Hara, Y. (1999). Antimutagenic and anticarcinogenic activity of tea polyphenols. Mutat. Res., 436: 69-97.

Kuroda, Y. (1996). Bio-antimutagenic effect activity of green tea catechins in cultured Chinese hamster V79 cells. Mutat Res., 361: 179-186.

Landau, J.M., Wang, Z.Y., Yang, G.Y., Ding, W. and Yang, C.S. (1998). Inhibition of spontaneous formation of lung tumors and rhabdomyosarcomas in A/J mice by black and green tea. Carcinogenesis, 19: 501-507.

Lee, S.F., Liang, Y.C. and Lin, J.K. (1995). Inhibition of 1,2,4-benzenetriol-generated active oxygen species and induction of phase II enzymes by green tea polyphenols. Chem. Biol. Interact., 98: 283-301.

Lee, I.P., Kim, V.H., Kang, M.H., Roberts, C., Shim, J.S. and Roh, J.K. (1997). Chemopreventive effect of green tea (*Camellia sinensis*) against cigarette smoke induced mutations (SCE) in humans. J. Cell. Biochem. Suppl., 27: 68-75.

Lee, S.F. and Lin, J.K. (1997). Inhibitory effects of phytopolyphenols on TPA-induced transformation, PKC activation, and C-jun expression in mouse fibroblasts cells. Nutr. Cancer, 28: 177-183.

Li, N., Sun, Z., Han, C. and Chen, J. (1999). The chemopreventive effects of tea on human oral precancerous mucosa lessions. Proc. Soc. Exp. Biol. Med., 220: 218-224.

Li, Y. (1991). Comparative study on the inhibitory effect of green tea, coffee and levamisole on the hepatocarcinogenic action of diethylnitrosoamine. Chinese J. Cancer, 13: 193-195.

Liang, Y.C., Lin-Shiau, S.Y., Chen, C.F. and Lin, J.K. (1997). Suppresssion of extracellular signals and cell proliferation through EGF receptor binding by (–)-epigallocatechin gallate in human A431 epidermal carcinoma cells. J. Cell. Biochem., 67: 55-65.

Liao, S., Kao, Y.H., Hippakka, R.A. (2001). Green tea: biochemical and biological basis for health benefits. Vitam. Horm., 62: 1-94.

Lin, J.K., Liang, Y.C. and Lin-Shiau, S.Y. (1999a). Cancer chemoprevention by tea polyphenols through mitotic signal transduction blockage. Biochem. Pharmacol., 58: 911-915.

Lin, W.W., Chang, S.H. and Wang, S.M. (1999b). Roles of a typical protein kinase in lysophosphatidic acid-induced type II adenyiyl cyclase activation in Raw 264.7 macrophages. Br. J. Pharmacol., 128: 1189-1198.

Lin, J.K. and Liang, Y.C. (2000) Cancer chemoprevention by tea polyphenols. Proc. Natl. Sci. Coun. ROC (B), 24: 1-13.

Lin, J.K. and Lee, S.F. (1995). Inhibition of tumor promotion through blocking mitotic signal transduction. Zool. Stud., 34: 67-81.

Lin, Y.L. and Lin, J.K. (1997). (–)-Epigallocatechin-3-gallate blocks induction of nitric oxide synthetase by downregulating lipopolysaccharide-induced activity of transcription factor nuclear factor-kB. Mol. Pharmocol., 52: 465-472.

Lin, Y.L., Cheng, C.Y., Lin, Y.P., Lau, Y.W., Juan, I.M. and Lin, J.K. (1998). Hypolipidemic effect of green tea leaves through induction of phase II enzymes including superoxide dismutase, catalase and glutathione-S-transferase in rats. J. Agric. Food. Chem., 46: 1893-1899.

Lin, Y.L., Tsai, S.H., Lin-Shiau, S.Y. Ho, C.T. and Lin, J.K. (1999). Theaflavin-3'3-digallate from black tea blocks the nitric oxide synthase by down-regulating the activation of NF-KB in macrophages. Eur. J. Pharmacol. 367: 379-388.

Liu, L. and Castonguay, A. (1991). Inhibition of metabolism and genotoxicity of 4-(methylnitrosoamino)-1-(3-pyridyl)-1-butanone (NNK) in rat hepatocytes by (+)-catechin. Carcinogenesis, 12: 1203-1208.

Liu, L., Han, C. and Chen, J. (1998). Short term screening of anticarcinogenic ingredients of tea in cell biology assays. Wei. Sheng. Yan. Jiu., 27: 53-56.

Liu, X.L. (1990). Genotoxicity of fried fish extract, MeIQ and inhibition by green tea antioxidant. Zhonghua Zhong Liu Za Zhi, 12: 170-173.

Lu Y.P., Lou Y.R., Xie J.G., Yen P., Huang M.T. and Conney, A.H. (1997). Inhibitory effect of black tea on the growth of established skin tumors in mice: effects on tumor size, apoptosis, mutosis and bromodeoxyuridine incorporation into DNA. Carcinogenesis 18: 2163-2169.

Maliakal, P.P., Coville, P.F., and Wanwimolrukh, S. (2001). Tea consumption modulates hepatic drug metabolizing enzymes. J. Pharm. Pharmocol., 53: 569-577.

Mao, R. (1993). The inhibitory effects of epicatechin complex on the diethylnitrosomine induced initiation of hepatocarcinogenesis in rats. Chinese J. Cancer, 27: 201-204.

Matsumoto, N., Kohir T., Okushio K. and Hara, Y. (1996). Inhibitory effects of tea catechins, black tea extract and oolong tea extract on hepatocarcinogensis in rat. Japan J. Cancer Res. 87: 1034-1038.

McKay, D.L. and Blumberg, J.B. (2002). The role of tea in human health: an update. J. Am. Coll. Nutr., 21: 1-3.

Miller, N.J., Castelluccio, C., Tijiburg, L. and Rice-Evans, C. (1996). The antioxidant properties of theaflavins and their gallate ester-radical scavengers or metal chelators. FEBS Lett., 392: 40-44.

Mimoto, J., Kiura, K., Matsuo, K., Yoshino, T., Takata, I., Ueoka, H., Kataoka, M. and Harada, M. (2000). (–)-Epigallocatechin gallate can prevent cisplantin-induced lung tumorogenesis in A/J mice. Carcinogenesis, 21: 915-919.

Moncada, S.R., Palmer, M.J. and Higgs, D.A. (1992). Nitric oxide: pathophysiology and pharmacology. Pharmacol., Rev., 43: 109-142.

Morse, M.A., Kresty, L.A., Steele, V.E., Kelloff, G.J., Boone, C.W., Balentine, D.A., Harbowy, M.E. and Stoner, G.D. (1997). Effects of theaflavins on N-nitrosomethylbenzylamine-induced esophageal tumorogenesis. Nutr. Cancer, 29: 7-12.

Nakamura, Y., Hara, Y. and Osawa, T. (2002). Black Tea Polyphenols, Theaflavins, Prevent Cellular DNA Damage by Inhibiting Oxidative Stress and Suppressing Cytochrome P450 1A1 in Cell Cultures. J. Agric. Food Chem., 50: 213-220.

Nakayama, M., Suzuki, K., Toda, M., Okubo, S., Hara, Y. and Shimamura, T. (1993). Inhibition of the intectivity of influenza virus by tea polyphenols. Antiviral Res. 21: 289-299.

Nishida, H., Omori, M., Fukutomi, Y., Ninomiya, M., Nishiwaki, S., Suganuma, M., Moriwaki, Y. and Muto, Y. (1994). Inhibitory effects of (–)-epigallocatechin gallate on spontaneous hepatoma in C3H/NeHCrj mice and and human hepatoma-derived PLC/PRF/5 cells. Japan. J. Cancer Res., 85: 221-225.

Nomura, M., Ma, W.Y., Huang, C., Yang, C.S., Bowden, G.T., Miyamoto, K. and Dong, Z. (2000). Inhibition of ultraviolet B-induced AP-1 activation by theaflavins from black tea. Mol. Carcinogen., 28: 148-155.

Okabe, S., Suganuma, M., Hayashi, M., Sueoka, E., Komori, A. and Fujiki, H. (1997). Mechanism of growth inhibition by of human cancer cell PC-9 by tea polyphenols. Jpn. J. Cancer Res., 88: 639-643.

Okai, Y. and Higashi-Okai, K. (1997). Potent suppressing activity of the non-polyphenolic fraction of green tea (*Camellia sinensis*) against genotoxin-induced umu C gene expression in *Salmonella typhimurium* (TA 1535/psK 1002) association with pheophytins a and b. Cancer Lett., 120: 117-123.

Okubo, T., Tode, Y., Hara, T., and Shimanure, T. (1991). Antifungal and antimicrobial activities of tea extract and catechins. Jpn. J. Bacteriol., 46: 509-514.

Okuda, T., Mori, K. and Hayatsu, H. (1984). Inhibitory effect of tanins on direct-acting mutagens. Chem. Pharm. Bull., 32: 3755-3758.

Pannala, A.S., Rice-Evans, C.A., Halliwell, B. and Singh, S. (1997). Inhibition of peroxynitrite-mediated tyrosine nitration by catechin by catechin polyphenols. Biochem. Biophys. Res. Commun., 232: 164-168.

Park, A.M. and Dong, Z. (2003). Signal transduction pathways: targets for green and black tea polyphenols. J. Biochem. Mol. Biol., 36: 66-77.

Park, E.J. and Pezzuito, J.M. (2002). Botanical in cancer chemoprevention. Cancer Metastasis Rev., 21: 231 – 255.

Quin, G., Gopalan-Kriczky, P., Su, J., Ning, Y. and Lotliker, P.D. (1997). Inhibition of aflatoxin B_1 induced initiation of hepatocarcinogenesis in rats by green tea. Cancer Lett., 112: 149-154.

Rasheed, A. and Haider M. (1998). Antibacterial activity of *Camellia sinensis* extracts against dental caries. Arch. Pharm. Res., 21: 348-352.

Rogers, A.E., Hafer, L.J., Iskander, Y.S. and Yang, S. (1998). Black tea and mammary gland carcinogenesis by 7,12-dimethylbenz[a]anthracene in rats fed control or high fat diets. Carcinogenesis, 19: 1269-1273.

Sano, M., Takahashi, Y., Yoshino, K., Shimoi, K., Nakamura, Y., Tomita, I., Oguni, I. and Konomotor, H. (1995). Effect of tea (*Camellia sinensis* L.) on lipid peroxidation in rat liver and kidney: a comparison of green and black tea feeding. Biol. Pharmaceutical Bull., 18: 1006-1008.

Sarkar, A. and Bhaduri, A. (2001). Black tea as the powerful chemopreventor of reactive oxygen and nitrogen species: comparison with its individual catechin constituents and green tea. Biochem. Biophys. Res. Commun., 284: 173-178.

Sasaki, Y.F., Yamada, H., Shimoi, K., Kator, K. and Kinae, N. (1993). The clastogen suppressing effects of green tea, Po-Lei tea and Rooibos tea in CHO cells and mice. Mutat. Res., 286: 21-232.

Satoh, E., Ishi, T., Shimizu, Y., Sawamura, S. and Nishimura, M. (2001). Black tea extract, thearubigin fraction, counteracts the effect of tetanus toxin in mice. Exp. Biol. Med. (Maywood)., 226: 577-580.

Sazuka, M., Murakami, S., Lsemura, M., Satoh, K. and Nukiwa, T. (1995). Inhibitory effects of green tea infusion on in vitro invasion and in vivo metastasis of mouse lung carcinoma cells by green tea and black tea in man. Eur. J. Clin. Nutr., 50: 28-32.

Serafini, M., Ghiselli, A. and Ferro-Luzzi, A. (1996). In vivo antioxidant effect of green and black tea in man. Eur. Clin. Nutr., 50: 28-32.

Shetty, M., Subbannaya, K. and Shivananda, P.G. (1994). Antibacterial activity of tea (*Camellia sinensis*) and coffee (*Coffee arabica*) with special reference to *Salmonella typhimurium*. J. Commun. Dis., 26: 147-150.

Shi, S.T., Wang, Z.Y., Theresa, J.S., J.S., Hong, S.Y., Chen, W.F., Ho, C.T. and Yang, C.S. (1994). Effect of green tea and black tea on 4-(methylnitrosamino)-1-(3-pyridyl)-1-butanone bioactivation, DNA methylation and lung tumorigenesis in A/J mice. Cancer Res., 54: 4641-4647.

Shimoi, K., Nakamura, Y., Tomita, I., Hara, Y. and Kaka, T. (1986). The pyrogallol related compounds reduce UV-induced mutations in *Escherichia coli* B/R WP2. Mutat. Res. 173: 239-244.

Shiraki M., Hara Y., Osawa T., Kumon H., Nakayama T. and Kawakishi S. (1994). Antioxidative and antimutagens effects of theaflavins from black tea. Mutat. Res. 323: 29-34.

Shukla, Y. and Taneja P. (2002). Anticarcinogenic effect of black tea on pulmonary tumors in Swiss albino mice. Cancer Lett., 176: 137-141.

Sohn, O.S., Surace, E.S., Richie, J.P.J., Colosimo, S., Zang, E. and Weisburger, J.H. (1994). Effects of green and black tea on hepatic xenobiotic metabolizing system in the male F344 rat. Xenobiotica, 24: 119-127.

Stravic, B., Matula, T.I., Klassen, R. and Downie, R.H. (1996). The effects of teas on the in vitro mutagenic potential of heterocyclic aromatic amines. Food. Chem. Toxicol. 34: 515-523.

Suganama, M., Okabe, S., Kai, Y. et al. (1999). Synergestic effects of (–)-epigallocatechin gallate with (–)-epicatechin, sulindac or tamoxifen on cancer preventive activity in human cancer cell line PC-9. Cancer Res., 59: 44-47.

Sur-Altiner, D. and Yenice, B. (2000a). Effect of black tea on lipid peroxide and glutathione levels in female rats. Drug. Metabol. Interact., 16: 299-305.

Tanaka, H., Hirose, M., Kawabe, M., Sano, M., Takesada, Y., Hagiwara, A. and Shirai, T. (1997). Post-initiation inhibitory effects of green tea catechins on 7,12-dimethylbenz-[a]anthracene-induced mammary gland carcinogenesis in female Sprague-Dawley rats. Cancer Lett., 116: 47-52.

Taneja, P. (2002). Ph.D. Thesis entitled 'Studies on antimutagenic potential of black tea and its constituents'. Jamia Hamdard, Hamdard University, New Delhi.

Toda, M., Okubo, S., Ikigai, H. and Shimamura, T. (1990). Antibacterial and anti-hemolysin activities of tea catechins and their structural relatives. Nippon Saikingaku Zasshi, 45: 561-566.

Toda, M., Okubo, S., Ohnishi, R. and Shimamura, T. (1989). Antibacterial and bactericidal activities of Japanese green tea. Nippon Saikingaku Zasshi, 44: 669-72.

Tomita, I., Sano, M., Sasaki, K. and Miyase, T. (1997). Tea catechin (EGCG) and its metabolites as bioantioxidants. In: Functional Foods for Disease Prevention I: Fruits, Vegetables and Teas, 21, 209-216. (Shibamoto, T., Terao, J. and Osawa, T., eds). ACS Symposium 701, Atlanta, USA.

Touyz, L.Z. and Amsel, R. (2001). Anticariogenic effects of black tea (*Camellia sinensis*) in caries prone-rats. Quintessence Int., 32: 647-50.

Valcic, S., Timmermann, B.N., alberts, D.S., Wachter, G.A., Krutzsch, M., Wymer, J. and Guillen, J.M. (1996). Inhibitory effect of six green tea catechins and caffeine on the growth of four selected human cell lines. Anticancer Drugs, 7: 461-467.

Vinson, J.A. (2000). Black and green tea and heart disease: a review. Biofactors, 13: 127-132.

Wang, Z.Y., Hong, J.Y., Huang, M.T., Reuhl, K.R., Conney, A.H., and Yang, C.S. (1992a). Inhibition of N-nitrosodiethylamine and 4-(methylnitrosoamine)-1-(3-pyridyl)-1-butanone-induced tumorigenesis in A/J mice by green tea and black tea. Cancer Res., 52: 1943-1947.

Wang, Z.Y., Huang, M.T., Lou, Y.R., Xie, J.G., Reuhl, K.R., Newmark, H.J.L., Ho, C.T., Yang, C.S. and Conney, A.H. (1994). Inhibitory effects of black tea, decaffeinated black tea, and decaffeinated green tea on ultraviolet B light-induced skin carcinogenesis in 7,12-dimethylbenz[a]anthracene-induced SKH-1 mice. Cancer Res., 54: 3428-3435.

Wang, Z.Y., Wang, L.D., Lee, M.J., Ho, C.T. Huang, M.T, Conney, A.H. and Yang, C.S. (1995). Inhibition of N-nitrosomethylbenzylamine induced oesophageal tumorigenesis in rats by green and black tea. Carcinogenesis, 16: 2143-2148.

Wang, Z.Y., Das, M., D.R. Bickers, D.R. and H. Mukhtar, H. (1988). Interaction of epicatechins derived from green tea with rat hepatic cytochrome P-450. Drug. Metab. Dispos. 16: 98-103.

Wang, Z-Y., Huang, M.T., Ferraro, T., Wong, C.Q., Lou, Y.-R., Reuhl, K., Iatropoulos, M., Yang, C.S. and Conney, A.H. (1992b). Inhibitory effects of green tea in drinking water on tumorogenesis by ultraviolet light and 12-O-tetradeconyl-13-acetate in skin of SKH-1 mice. Cancer Res., 52: 1162-1170.

Weisburger, J.H. (1999). Tea and health: the underlying mechanisms. Proc. Soc. Exp. Biol. Med., 220: 271-275.

Weisburger, J.H., Hara, Y., Dolan, L., Luo, F.Q., Pittman, B. and Zang, E. (1996). Tea polyphenols as inhibitors of mutagenicity of major classes of carcinogens. Mutat. Res., 371: 57-63.

Xu, M., Bailey, A.C., Hernaez, J.F., Taoka, C.R., Schut, H.A. and Dashwood, R.H. (1996). Protection by green tea, black tea, and indole-3-carbinol against 2-amino-3-methylimidazo[4,5-f]quinoline-induced DNA adducts and colonic aberrant crypts in the F344 rat. Carcinogenesis, 17: 1429-1434.

Xu, Y., Ho, C-T., Amin, S.G., Han, C. and Chung, F-L. (1992). Inhibition of tobacco-specific nitrosoamine-induced lung tumorogenesis in A/J mice by green tea and its major polyphenols as antioxidants. Cancer Res., 52: 3875-3879.

Yamaguchi, K., Honda, M., Ikigai, H., Hara, Y. and Shimamura, T. (2002). Inhibitory effects of (–)-epigallocatechin gallate on the life cycle of human immunodeficiency virus type 1 (HIV-1). Antiviral Res., 53: 19-34.

Yanagawa, Y., Yamamoto, Y., Hara, Y., and Shimamura, T. (2003). A combination effect of epigallocatechin gallate, a major compound of green tea catechins, with antibiotics on *Helicobacter pylori* growth in vitro. Curr. Microbiol., 47: 244-249.

Yang, C.S, Maliakal, P., and Meng, X. (2002). Inhibition of carcinogenesis by tea. Annu. Rev. Pharmacol. Toxicol., 42: 25-54.

Yang, C.S., Prabhu, S., and Landau, J. (2001). Prevention of carcinogenesis by tea polyphenols. Drug Metab. Rev., 33: 237-253.

Yang, C.S., Chung, J.Y., Yang, G.Y., Li, C., Meng, X. and Lee, M.J. (2000a). Mechanisms of inhibition of carcinogenesis by tea. Biofactors, 13: 73-79.

Yang, G.Y., Liao, J., Li, C., Chung, J., Yurkow, E.J., Ho, C.T. and Yang, C.S. (2000b). Effect of black and green tea polyphenols on c-jun phosphorylation and H_2O_2 production in transformed and non-transformed human bronchial cell lines: possible mechanisms of cell growth inhibition and apoptosis induction. Carcinogenesis, 21: 2035-2039.

Yang, G.Y., Liao, J., Kim, K., Yurkow, E.J. and Yang, C.S. (1998). Inhibition of growth and induction of apoptosis in human cell lines by tea polyphenols. Carcinogenesis, 19: 611-616.

Yeh, C.W., Chen, W.J., Chiang, C.T., Lin-Shiau, S.Y., and Lin, J.K. (2003). Suppression of fatty acid synthase in MCF-7 breast cancer cells by tea and tea polyphenols: a possible mechanism for their hypolipidemic effects. Pharmacogenomics J., 3: 267-76.

Yokozawa, T., Cho, E.J. Hara, Y. and Kitani, K. (2000). Antioxidant activity of green tea treated with radicle initiator 22'Azobis (2-amidino propane) dihydrochloride. J. Agric. Food. Chem., 48: 5068-5073.

Yoshizawa, S., Horiuchi, T., Fujiki, H. et al. (1987). Antitumor promoting activity of (–)-epigallocatechin gallate, main constituent of " tannin" in green tea. Phytotherp. Res., 1: 44-47.

Zao, J.Y., Zhu, Q.F. and Cheng, S.J. (1992). The antagonistic action effects of green tea on micronuclei and apoptosis induced by 1,2-dimethylhydrazine (1,2 DMH) in colonic crypt cells of mice. Acta. Nutr. Sin., 14: 251-259.

Zhang, H., Spitz, M.R., Tomlinson, G.E., Schabath, M.B., Minna, J.D. and Wu, X. (2002). Modification of lung cancer susceptibility by green tea extract as measured by the comet assay. Cancer Detect. Prev., 26: 411-418.

Zhang, A., Zhu, Q.Y., Luk, Y.S., Ho, K.Y., Fung, K.P. and Chen, Z.Y. (1997). Inhibitory effects of jasmine green tea epicatechin isomers on free radical-induced lysis of red blood cells. Life Sci., 61: 383-394.

Zhang, G., Miura, Y. and Yagasaki, K. (2000). Induction of apoptosis and cell cycle arrest in cancer cells by in vivo metabolites of teas. Nutr. Cancer., 38: 265-273.

Zvetkova, E., Wirleitner, B., Tram, N.T., Schennachm, H. and Fuchs, D. (2001). Aqueous extracts of *Crinum latifolium* (L.) and *Camellia sinensis* show immunomodulatory properties in human peripheral blood mononuclear cells. Int. Immunopharmocol., 1: 2143-2150.

Chapter

6

Tea as a Protectant in Human Cancer

Geeta Talukder[*] and Archana Sharma[**]
[*]Senior Scientist and Consultant
Vivekananda Institute of Medical Sciences, Kolkata, India
[**]INSA Honorary Scientist
Centre of Advanced Studies in Cell and Chromosome Research
Department of Botany, University of Calcutta, Kolkata, India
e mail: geetatalukdar@hotmail.com, geatatalukder@yahoo.co.in

INTRODUCTION

Tea has been used in India for many centuries. Called *Shyamaparni* (Black leaf) and *Atandri* (preventor of sleep) as it is known in Hindi, was an essential ingredient in more than 30 ayurvedic preparations. Amongst people living in the hills of Assam, Kashmir, Tibet and Nepal, tea has been traditionally taking as infusion with hot water (called locally fant) with no sugar or milk, as a remedy for various ailments (Lal and Sen, 1903; Bhattacharya, 1989).

Tea was considered a mental stimulant, mood elevator, and stimulator of throat and bronchial secretion and was advocated for sore throat, chronic cough and oral ulcers. Unfortunately, the effects of tea on cancer have not been documented.

Later, during the 19th and 20th century, when irritant factors were considered to be initiators, or promoters of cancer, factors like betel leaf with tobacco quid chewing, cigarette smoking, alcohol intake and bad oral hygiene were said to be the main factors responsible for the high incidence of oral cancer (OC) in India. It was, however, observed that betel leaf alone or with lime did not induce carcinomas; it was the addition of tobacco and betel nut that caused the irritant action. Though the population on which the study was done may have also taken tea, this factor was not considered until tea, specially green tea, was reported to be a carcinopreventive agent in China and Japan. The reason is attributed mainly to the fact that malignant transformation is governed by genetic and epigenetic factors. It is a multifactorial condition and it is difficult to carry out epidemiological studies with accuracy in heterogenous subjects like man. It is also quite possible that the incidence of oral cancer achieved noticeable levels only with the advent of the tobacco habit after the 16th century. WHO has estimated that by the year 2015 there will be 15 million new cancer cases annually and approximately two-thirds will be in the developing countries (Dinshaw et al., 2003). Prevention depends on basic researches or cancer biology and heterogeneity (Sen Pathak, 2003). Oral cancer (OC) is specially important as it is the foremost among all cancers in males and third most common in females in India (Parkin et al., 1997, ICMR, 1992). Sankaranarayanan et al., (1992) reported that 7% of all cancer deaths in males and 4% in females are due to OC.

Lifestyle and Diet in Cancer

Epidemiological surveys and experimental studies have provided evidence that environmental factors, including dietary substances, play a major role in the development of cancer (ACS, 1995, Cao et al., 1996; Deakin et al., 1996, Enwonwu and Meeks, 1995; Epstein, 1993; Lin and Liang, 2000). The mechanism of carcinogenesis involves multiple stages of biochemical and molecular alterations in target cells. The process of skin carcinogenesis involves stepwise accumulation of genetic changes, ultimately leading to malignancy (DiGiovanni, 1992; Slaga et al., 1995, 1996). There are three main steps: (1) The first step in multistage skin carcinogenesis is initiation, which involves carcinogen-induced genetic changes. The changes at this stage appear to be rapid and irreversible. It is presumed to involve irreversible modification of DNA, perhaps resulting in one or more mutations (Garner, 1998). (2) A much slower stage of carcinogenesis is promotion, which is believed to be involved in selective and sustained hyperplasia, leading to the specific expansion of initiated cells into papillomas or to precancerous lesions (Slaga et al., 1996). These changes may result from epigenetic mechanisms, such as action of ornithine decarboxylase (ODC), protein kinase C (PKC) and prostaglandins. (3) The process of tumour progression is characterized by a high level of genetic instability that leads to a number of chromosomal alterations (Warren et al., 1993, Sen Pathak, 2003). Epigenetic factors like dietary habits, tobacco and alcohol intake may also promote preneoplasia and neoplasia in man.

The development of cancer is, therefore, a multistep process involving sequential initiation, promotion, transformation and progression of the disease. During initiation/mutagenesis, the effect is on a critical DNA segment. The latter may be a *proto-oncogene,* which regulates cell proliferation and differentiation, altering to an oncogene, or a tumor suppressor gene (which becomes non-functional as a suppressor). In both cases, the cells will proliferate, freed from regulatory controls.

Nutritional influences act not only on the initiation and promotion levels but also influence on the host-immune response. Cancer cells escape immune destruction and grow undisturbed by suppressing the immune response or by not expressing appropriate membrane markers required for immune recognition and destruction (Sanders and Kline, 1995). It has also been suggested that tumour metastasis and spread are related to dietary and other nutritional factors and differences in dietary fibre intake.

The relationship of different dietary and other lifestyle factors with incidence of cancer in population surveys tends to be over-simplified. Humans are exposed to complex adulterations in the environment, including diet etc.

Betel chewing was considered a positive factor for increased OC in India (Notani, 2000). Specifically, the main ingredients of betel quid, viz., betel nut, tobacco, calcium and catechu, besides other condiments, form a complex mixture. Sen et al., (1986, 1989, 1994), while screening and testing the different ingredients on mice with unscheduled DNA synthesis of the gastric mucosa, had shown that tobacco consumed alone exerted the maximum carcinogenic effect (Bandyopadhyay and Sharma, 1980). The betel leaf, on the other hand, did not increase unscheduled DNA synthesis and appeared actually to resist or reduce the carcinogenic effects of tobacco and betel nut. Irritant effects of tobacco on the oral and gastric mucosa have also been observed in man, using micronuclei and other tests.

Nitrosamines, administered directly in the diet or induced *in vivo,* have been known to be the prime causes of gastric and oesophageal cancer. Evidence from pathological and epidemiological studies suggests that nitrosation can originate in the gastric mucosa, gastric cavity or inflammatory infiltrate (Correa, 1992; Mirvish, 1986). *N-nitroso* compounds are formed due to reaction of nitrates with amides and amines in the diet in the presence of gastric and salivary secretions. Nitrates are found in amounts about

100 mg/day in normal diet and can be converted to nitrite. Nitrites are also present in additives and preservatives. Primary amines are found in cheese, fish, pickles and preserved meat. The risk factor of amino acids present in meat, milk, cheese and eggs are relatively negligible (DeFlora and Ramel, 1988; De Flora et al., 1992).

Effect of Tea in Animal Systems

The effect of tea has been studied mostly in the oral and gastrointestinal tract of animals for direct effect and in the bacterial and animal test systems for mutagenicity. Hydroxylated flavonoids in tea had been reported to have a potent inhibitory effect on mutagenic activity and prevented DMBA-induced skin carcinogenesis in mice (Yoshizawa et al., 1987; Yang and Wang, 1993). Topical application and oral feeding of green tea polyphenols have been shown to inhibit 3-methylchlorathrene or UV-induced cancers (Wang et al., 1991, 1994; Gensler et al., 1996). Others have shown that non-polyphenol fraction of green tea inhibited tumor formation in mouse skin, as well as benzo (a) pyrene induced for stomach cancers (Katiyar et al., 1993; Wang et al., 1992). Apart from skin and gastrointestinal cancers, green and black tea were found to decrease the spontaneous formation of lung tumors and rhabdomyosarcomas in A/J mice and in azoxymethane induced colon carcinogens in Fisher rats (Yamane et al., 1991).

Green, black and decaffeinated black tea were found to significantly inhibit the mutagenicity induced by three indirect-acting dietary carcinogens, Glu-p-l.B[a]P and nitrosopyrrolidine, in a bacterial test system (Bu-Abbas et al., 1996). Tea polyphenols sharply decreased the mutagenicity of a number of aryl- and heterocyclic amines, of aflatoxin B1, B[a]P, 1,2-dibromoethane, and more selectively, of 2-nitropropane in *Salmonella typhimurium* (Weisburger et al., 1996). Yen and Chen (1996) studied the relationship between the chemical content and antimutagenic activity of various tea extracts. The amount of catechins in various tea extracts was in the order: green tea > oolong tea > pouchong tea > black tea. The tea extracts of oolong and pouchong tea markedly inhibited the mutagenic activity of various carcinogens. Recently, we also examined the antimutagenic properties of various tea extracts (green, pouchong, oolong and black tea) against different groups of environmental mutagens. The results suggested that various degrees of tea fermentation might produce diverse types of antimutagenic compounds effective against their corresponding mutagens (Hour et al., 1999). Fumes from cooking oils were found to have mutagenic activity (Wu et al., 1998), indicating that nitro- and polycyclic aromatic hydrocarbons (PAHs), which may be a risk factor for lung cancer among cooks, could be reduced by adding catechin. Polyphenols in black tea polyphenols were more potent inhibitors of mutagenicity than green tea in ones caused by the food mutagen 2-amino-1-methyl-6-phenylimidazo [4,5-b] pyridine (PhIP) and other mutagens in the *Salmonella* test system (Apostolides et al., 1996, Chen and Ho, 1994; Chen and Yen, 1997). Hernaez et al., (1998) also suggested that catechin might protect against such diverse reactive intermediates as free radicals and electrophiles formed during the metabolic activation of IQ. Other workers have shown inhibitory action of tea on cultured mammalian cells (Imanishi et al., 1991) and on animals treated with mutagens *in vivo* (Mukherjee et al., 1999).

Epidemiological studies worldwide, using food frequency questionnaires, had implicated dietary and nutritional factors in the development of oral and pharyngeal cancers. Staple diet consisting of poor calories and micronutrients, along with supplementation with chilli powder and wood smoke used for cooking, have been linked to oral leukoplakia and primary oral cancer. Japanese populations had shown lower incidence of colonic cancer after immigration to the West, which may be related to altered life style (Doll and Peto, 1981). Exposure to a caricinogen can be modified by other factors which are again influenced by diet, as for example, microflora of the gut. Various factors in the lifestyle may interact at many levels to alter the balance of activation and deactivation of carcinogens after entry.

The site of cancer also appears to be of importance. Easily available sites, such as the oesophagus, stomach and lungs are more susceptible to induction of cancer by external agents than organs like kidney, liver and haematopoeitic systems. The inhibiting mechanism act at different levels. Wattenberg (1985) had given a broad range of chemicals known to be inhibitors. De Flora and Ramel (1988) further classified the inhibitors of mutagenesis and carcinogenesis into three broad categories: a) those acting extra-cellularly, b) those acting intra-cellularly and c) those modulating progression.

Prevention of Cancer in Man

Cancers of accessible sites like oral cavity, oesophagus, prostate and colon have been used for studying protection afforded by dietary agents (Anderson and Conning, 1992; Bruce, 1990; Ioannides, 1994). Stich et al., (1991a, 1991b) have reported remission of oral precancerous lesions of tobacco and areca nut chewers following administration of B-carotene or vitamin A in human populations at different geographical locations (Guam, Peru, Taiwan, Philippines and India).

Vitamin A, given at level of 50000 IU/week, was able to keep the frequency of micronucleated mucosal cells at low levels for at least a 12-month post-treatment period. Prevention of cancer of oral cavity and pharynx had been studied by earlier workers who had suggested abolition of the tobacco chewing habit and intake of vitamin A and retinoic acid (Baden, 1987; Franco et al., 1989; Jetten et al., 1990).

Cancer chemoprevention can be defined as the prevention, inhibition, or reversal of carcinogenesis by administration of one or more chemical entities, either as individual drugs or as naturally occurring constituents of the diet. Flavonoids are naturally occurring low molecular weight polyphenolic compounds widely distributed in fruits, vegetables and beverages. People who eat diets rich in fruits and vegetables have a lower incidence of diseases such as cancer (Cao et al., 1996; Sharma and Talukder, 2000; Ziegler, 1991). Numerous experimental studies have examined the specific role of flavonoids in disease prevention. For example, increased flavonoid intake was associated with decreased risk of carcinogenesis (DiGiovanni, 1992; Slaga et al., 1995, 1996).

The inhibitory effects of tea against carcinogenesis have been attributed to the biological activities of the polyphenol fractions in tea (Yang et al., 1997; Yen et al., 1997; IARC, 1991).

EFFECTS OF TEA ON THE INCIDENCE OF HUMAN CANCER

The effects of tea consumption on the occurrence of human cancers have been reviewed by the International Agency for Research on Cancer Working Group based on the literature available in 1989 (IARC, 1991) and by Yang and Wang based on the literature up to 1992 (Yang and Wang, 1993). A general conclusion regarding the relationship between tea consumption and cancer risk based on these studies is, however, difficult to draw.

Among 100 epidemiology studies, on the relationship, between tea drinking and human cancer, 28 studies showed no relationship and 14 studies showed a negative association between tea drinking and cancer incidence (Yang and Wang, 1993).

Experimental studies were conducted globally by Dreosti et al. (1997) on humans suffering from various forms of cancer and who consumed tea on a regular basis. Most of the studies which showed no association or were positive, had been carried out in Western countries, whereas a negative association was mostly found in case control studies in Asian countries, especially China and Japan, where inhabitants drank large quantities of green tea or oolong tea daily. A comparison of age-specific cancer incidence rates, a comparison of ages at cancer onset among cancer patients with different levels of consumption of green tea, a comparison of age- standardized cancer incidence rates and examination of the relative risk adjusted for other lifestyle population. They thus found a negative association between green tea consumption and cancer incidence, especially among females drinking more than 10 cups a day. A gradual decrease of cancer incidence with age was observed among females who consumed more than 10 cups a day; consumption of green tea was associated with late onset of cancer. The age-standardized average annual incidence rate was significantly lower among females who consumed large amounts of green tea. The relative risk of cancer incidence was also lower among both females and males in the highest tea consumption groups. This prospective cohort study showed that green tea had a potentially preventive effect against cancer in humans (Imai et al., 1997; Nakachi et al., 1998).

INTERACTION OF GENETIC ACTIONS IN CARCINOGENESIS IN HUMANS

Genetic factors play an important role in human cancer, development, especially in those related to epigenetic factors and explain individual susceptibility to environmental cancers as has been observed in population studies where members of same family equally exposed do not develop neoplasia. In the study of complex factors in cancer development, they are increasingly being studied. In oral and gastrointestinal cancers especially the enzyme polymorphism of cytochrome P450, CYP, glutathione S-transferase (GSTs) may be of importance and the relationship with, tobacco derived carcinogens has been studied in populations of various countries, i.e., India (Sreelekha, et al., 2001, Bhisey, et al., 2003, Roy and Sikdar, 2003), China (Hung et al., 1997; Zheng et al., 1996), Japan (Morita et al., 1999) and others (Loktiono et al., 1997, Oude-Ophius et al., 1998; Olshan et al., 2000; Park et al., 2000).

Squamous cell carcinoma of the head and neck (SCCHN) is a group of cancers of epithelial origin that may provide an ideal model for the study of gene–environment interaction. SCCHN includes squamous cell carcinomas of the oral cavity, pharynx and larynx. Approximately 90% of the attributable risk for oral cancer and 80% of the attributable risk for larynx cancer result from tobacco intake (Bartsch et al., 2000; Cheng et al., 1999; Cosma et al., 1993). It has been proved beyond doubt that tobacco smoking increases the risk of SCCHN in a dose-response manner. Polymorphisms of carcinogen-metabolizing enzymes, known to be involved in metabolism of carcinogens present in tobacco smoke, are relatively common in most populations. In a concise review of 24 published studies that evaluated the risk of SCCHN in relation to two "deletion" polymorphisms of the glutathione S-transferase family: GSTM1 and GSTT1, patterns of risk based on the site of the tumour and on community are presented, as are some methodological weaknesses. The results of these studies are inconsistent, with some reporting weak-to-moderate associations and others finding no elevation in risk for the main effect of the gene. Few studies have directly evaluated the interaction with tobacco or tea habits. Well designed, population-based comprehensive studies are needed covering all aspects of diet (which should include alcohol, tea and coffee) as well as different constituents of betel quid. Like all dietary studies, it is apparent that a balanced diet includes both protectants and promutagens, and also the fact that substances of plant origin are more effective in raw form than as individual chemical agents (Sharma and Talukder, 2000; Sarkar et al., 1994, 1996).

Zavras et al., 2002) investigated effects of OC risk of an interaction between a single nucleotide polymorphism (SNP) in the alcohol dehydrogenase 3 (ADH3) gene and alcohol consumption levels using a hospital-based study of 93 cases and 99 controls conducted in Athens (Greece). This SNP affected ethanol metabolism *in vitro* and appeared to interact with alcohol consumption in a previous OC study. They also evaluated a SNP in CYP2E1, another gene involved in ethanol metabolism, reported to be associated with OC risk in a European population. Data on genotypes and risk factors obtained from interviews were analyzed using multivariate logistic regression, accounting for potential confounders. No overall (marginal) association was found between OC risk and ADH3 genotypes. An interaction between ADH3 genoptypes and alcohol consumption levels, however, was suggested. In non-drinkers, the ADH3(1-1) genotype has higher risk than ADH3(1-2) or ADH3(2-2) genotypes, but for subjects consuming alcohol, lower risk was observed for ADH3. They fit a logistic regression model to estimate the increase in OC risk associated with each alcohol drink consumed per week. They estimated that OC risk increased by 31.5% per drink per week for the ADH3(2-2) genotype, 4.1% for the ADH3 (1-2) genotype and 1.6% for the ADH3(1-1) genotype. Evidence of genotype-environment interaction was suggestive (Wald chi p = 0.145). This finding is contrary to that reported for a population in Puerto Rico, where the ADH3(1-1) genotype seemed more sensitive to ethanol exposure. In Greece, genetic variation at the CYP2E1 SNP is almost entirely absent, with only 1 case and 1 control heterozygous for the variant. By contrast, in a population in France where an OC association was reported, the frequency of CYP2E1 heterozygotes was 5% in controls and 9X in OC cases. These findings illustrate the importance of replicating SNP associations both within and between different racial and ethnic groups and geographic regions.

Sreelekha et al., (2001) studied genetic polymorphisms in CYP1A1, GSTM1 and GSTT1 genes in OC patients from Kerala. The study included 98 cases and 60 healthy controls which were age and sex matched. Genotypes of CYP1A1, GSTM1 and GSTT1 were determined by PCR-RFLP. For GSTM1 null deletion was observed in 49% of oral cancer cases and 33% of control subjects. For GSTT1, 18% of carcinomas and 8% of controls had the null genotype. In the case of CYP1A1 m2 allele, 51% of oral cancers and 17% of normal controls, respectively, had one or both alleles with the isoleucine -> valine substitution. Digestion of the PCR products with enzyme Nco1 revealed polymorphism for CYP1A1 m2 with bands at 263 bp. There was no association between genotypes with tumour size, stage, grade, and age. Since null genotype individuals may possibly be poor detoxifiers with reduced ability to neutralize the reactive carcinogenic intermediates, they may be a high risk category. The frequency distribution of CYP1A1 m2 (lle/val) genotypes among oral cancer patients was significantly different that from normal controls. The risk of GYP1A1 can be supported by the functional difference between presence of valine and isoleucine; valine type has higher catalytic and mutagenic activity towards benzo(a] pyrene than the isoleucine type. The results suggested that polymorphism in CYP1A1 m2 gene and/or GSTM1 and GSTT1 null genotype could confer an increased risk for oral cancer. The lifestyle habits do not mention coffee and tea intake. Polymorphism of the CYP and GST gene loci has also been studied in tobacco-related cancers (Bhisey et al., 2003). The studies open up new avenues for abundant experimental evidence of OC related to genotypes.

CONCLUSION

Although animal and bacterial mutagenicity studies have shown a definite protectant effect of tea both in raw form as well as individual constituents, effect on cancer in man is not yet unequivocal. However, with genotype-phenotype studies being initiated, it would be a matter of time before studies on tea drinking habits related to susceptible genotypes to cancer will yield useful results. Thus, tea taken regularly in any form, appears to be a very healthy drink, and the rich antioxidants levels are likely to be of benefit in the inherent protective system in man against carcinogens especially affecting the upper gastrointestinal tract and mouth.

REFERENCES

American Cancer Society, (ACS) (1995). Cancer factors and figures. pp. 1-3, Atlanta GA, USA.

Anderson, D. and Conning, D.M. eds. (1992). Experimental Toxicology: The Basic Principles. New Castle Atheneum Press.

Apostolides, Z., Balentine, D.A., Harbowy, M.E. and Weisburger, J.H. (1996). Inhibition of 2 amino-1-methyl-6 phenyl imidazo (4,5) pyridine mutagenicity by black and green tea extracts and polyphenols. Mutat. Res., 359: 159-163.

Baden, E. (1987). Prevention of cancer of the oral cavity and pharynx. *Cancer J. Clin.*, 37(1): 49-62.

Bandyopadhyay, A. and Sharma, A. (1980). Tobacco and its mutagenic effects. The Nucleus, 23(3): 157-169.

Bartsch, H., Risch, A., Rajas M., Wikman, H. and Alexandrov, K. (2000). Genetic polymorphism of CYP genes alone or in combination as a risk modifier of tobacco related cancers. Cancer Epidemiol Biomarkers Prevent., 9: 3-28.

Bhattacharya, S.K. (1989). Chiranjeeb Banoushadhi. Vol. 10, Ananda Publishers, Calcutta. pp. 100-109.

Bhisey, R., Kotekar, A. and Buch, S. (2003). Polymorphism at CYP and GST gene loci and susceptibility to tobacco related cancers. Proc. Ind. Nat. Sci. Acad. (8), 69(1): 35-48.

Bruce, W.R. (1990). Short term tests and long term prospects for colon cancer prevention. Med. Oncol. Tumor. Pharm., 7: 131-136.

Bu-Abbas, A., Nunez, X., Clifford, M.N., Walker, R. and Ioannides, C.A. (1996). Comparison of antimutagenic potential of green, black and decaffeinated teas. Mutagenes, 11: 597-603.

Cao G.H., Sofic, E. and Prior, R.L. (1996). Antioxidant capacity of tea and common vegetables. J. Agr. and Food. Chem., 44: 3426-3431.

Cheng, L., Sturgis, E., Eicher, S., Char, D., Spitz, M. and Wei, Q. (1999). GSS polymorphism and risk of squamous cell carcinomas of head and neck. Int. J. Cancer, 84: 220-224.

Chen, C.W. and Ho, C.T. (1994). Antioxidant properties of polyphenols extracted from green and black teas. J. Food. Lipids., 2: 35-36.

Chen, H.Y. and Yen, G.C. (1997). Possible mechanisms of antimutagens by various teas as judged by their effects on mutagenesis by 2:3 methyl imidazole quinolone and benz(o) pyrene. Mutat Res., 393: 115-122.

Correa, P. (1992). Diet modification and gastric cancer prevention. Monogr. Nutr. Cancer. Inst., 12: 75-78.

Cosma, G., Croft, S.F., Currie, D., Toniolo, P. and Garte, S.J. (1993). The relationship between genotype and function of the CYP1Al gene. J. Toxicol. Environ. Hlth., 40: 315-322.

Deakin, M., Elder, J., Hendrickse, C., Pekham, D., Baldwin, D., Pantin, C., Wild, N., Leopard, P., Bell, D.A., Jones, P., Duncan, H., Brannigan, K., Aldersea, J., Fryer, A.A. and Strange, R.C. (1996). GSTT1 genotypes and susceptibility to cancer. Studies of interactions with GSTM1 in lung, oral, gastric and colorectal cancers. Carcinogens, 17: 881-884.

De Flora, S. and Ramel, C. (1988). Mechanisms of inhibitors of mutagenesis and carcinogenesis. Mutat. Res., 52: 285-306.

De Flora, S., Bronzetti, G- and Sobels, F.H. (1992). Assessment of antimutagenecity. Mutat. Res., 267: 153-155.

DiGiovanni, J. (1992). Multistage carcinogenesis in mouse skin. Pharmacol. Ther., 54: 63-128.

Dinshaw, K.A., Notani, P.N. and Sarin, R. (2003). Projecting oncology into next millennium. Proc. Ind. Nat. Sci. Acad. (B), 69(1): 1-10.

Doll, R. and Peto, R. (1981). The causes of cancer: quantitative estimates of risks of cancer in USA. J. Natl. Cancer. Inst., 66: 1191-1308.

Dreosti, I.E., Wargovich, M.J. and Yang, C.S. (1997). Inhibition of carcinogenesis by tea: the evidence from experimental studies. Crit. Rev. Food. Sci. and Nutr., 37: 8.

Enwonwu, C.O. and Meeks, V.I. (1995). Bionutrition and oral cancer in humans. Crit. Rev. Oral. Biol. Med., 6: 5-17.

Estein, S.S. (1993). Evaluation of national cancer program and proposed reforms. Int. J. Hlth. Serv., 23: 15-44.

Franco, E.L., Kowalski, L.P., Oliviera, B.V., Curado, M.P., Pereira, R.N., Silva, M.E., Fava, A.S. and Toricini, H. (1989). Risk factors for oral cancer in Brasil : a case control study. Int. J. Cancer., 43(6): 992-1000.

Garner, R.C. (1998). The role of DNA adducts in chemical carcinogenesis. Mutat. Res., 402: 67-75.

Gensler, H.L., Timmermann, B.N., Valcics, Wachter, G.A., Dorr, R., Dvorakova, K. and Alberts, D.S. (1996). Prevention of photocarcinogenesis by topical administration of pure epigallocatechin gallate isolated from green tea. Nutr. Cancer. 26: 325-335.

Gensler, S.A. and Olshan, A.F. (2001). GSTM1, GSTT1 and the risk of squamous cell carcinoma of the head and neck. Am. J. Epidemiol., 154(2): 95-105.

Hernaez, J.F., Xu, M. and Dashwood, R.H. (1998). Antimutagenic activity of tea towards 2-hydroxy-amino-3-methyl imidazo quinoline effect of tea concentration and brew time on electrophile scavenging. Mutant. Res., 402: 299-306.

Hour, T.C., Liang, Y.C., Chu, I.S., Lin, J.K. (1999). Inhibition of eleven mutagens by various tea extracts. Food. Chem. Toxicol. 37: 569-579.

Hung, H.C., Chuang, J., Chien, Y.C., Chem, H.D., Chiang, C.P., Kvoy, S., Hildesheim, A. and Chen, C.J. (1997). Genetic polymorphism of CYP2E1, GSTM1, GSTT1 environmental factors and risk of cancer. Epidemiol. Biomarkers. Prev., 6(11): 901-905.

IARC, (International Agency for Research on Cancer). (1991). Coffee, tea, mate, methylxanthine and methylglyoxal carcinogenic risk human IARC Monogr. Eval., 51: 1-513.

ICMR, (Indian Council of Medical Research). (1992). In: ICMR Biennial Report. Population based cancer registry 1988-89. New Delhi National Cancer Registry Programme.

Imanishi, H., Sasaki, Y.F., Ohta, T., Watanabe, M., Kato, T. and Shirasu, Y. (1991). Tea tannin components modify the induction of SCF and chromosome aberration in mutagen-treated cultured mammalian cells and mice. Mutat. Res., 259: 79-87.

Imai, K., Suga, K. and Nakachi, K. (1997). Cancer preventive effects of drinking green tea among a Japanese population. Prev. Med., 26: 769-775.

Ioannides, C. (1994). Drugs, diet and disease: mechanistic approaches to cancer. Vol. 1, Simon and Schuster Pub., New York.

Jetten, A.M., Kim, J.S., Sacks, P.G., Rearick, J.I., Lotan, D., Hong, W.K. and Lotan, R. (1990). Int. J. Cancer., 45(1): 195-202.

Katiyar, S.K., Agarwal, R. and Mukhtar, H. (1993). Protective effects of green tea polyphenols administered by oral intubation against chemical carcinogen induced for stomach neoplasia. Cancer. Lett., 73: 167-172.

Lal, B. and Sen, A. (1903). As qnoted by Bhattacharya (1989).

Lin, J.K. and Liang, Y.C. (2000). Cancer chemoprevention by tea polyphenols. J. Nat. Cancer. Res., 24(1): 1-13.

Loktiono, V., Bingham, S.A. and Cummings, J.H. (1997). Diet genotype interactions: an example from tea. Brit. J. Nutr., 78(6): 1043-1044.

Mirvish, S.S. (1986). Effects of vitamin C and vitamin E on N-nitroso compound formation. Carcinogens and Cancer, 58: 1842-1850.

Morita, S., Yano, M., Tsujinaka, T., Akiyama, Y., Taniguchi, M., Kaneko, K., Miki, H., Fujii, T., Yoshino, K., Kusuoka, H. and Monden, M. (1999), Genetic polymorphisms of drug metabolising enzymes and susceptibility to head and neck and squamous cell. Int. J. Cancer, 80(5): 685-688.

Mukherjee, P., Sarkar, D. and Sharma, A. (1997). Effects of dietary consumption of black tea infusion alone and in combination with known clastogens on mouse bone marrow chromosomes in vivo. Food and Chem. Tox., 35(7): 657-661.

Mukherjee, P., Poddar, S., Talukder, G. and Sharma, A. (1999). Protection by black tea extract against chromosome damage induced by two heavy metals in mice. Pharmac. Biol., 37(3): 243-247.

Nakachi, K., Suemasu, K., Suga, K., Takeo, T., Imai, K. and Higashi, Y. (1998). Influence of drinking green tea an breast cancer malignancies among Japanese patients. Jpn. J. Cancer Res., 89: 254-261.

Notani, P.N. (2000). Epidemiology and prevention of head and neck cancer: a global view. In: Contemporary Issues in Oral Cancer, ed. D. Saranath. pp. 1-29. New Delhi, Oxford University Press.

Olshan, A.F., Weissler, M.C., Watson, M.A. and Bell, D.A. (2000). GSTM1, GSTT1, GSTP1, CYPIA1 and NAT1 polymorphisms in tobacco use and the risk of head and neck cancer. Cancer Epidemiol. Biomarkers Prev., 9: 165-191.

Oude-Ophius, M.B., van Lieshout, E.M.M., Roelofs, H.M.J., Peters, W.H.M. and Manni, J.J. (1998). Glutathione S transferase M1 and T1 and cytochrome P450 1A1 polymorphisms in relation to the risk for benign and malignant head and neck lesions. Cancer, 82: 936-943.

Park, A., Muscat, I.Y.A., Kaur, J.E.A., Schantz, T.A., Stern, S.P.A., Richie, J.C.A. and lazarus, P.T. (2000). Comparison of GSTmu polymorphisms and risk for oral cancer between African Americans and Caucasians. Pharmacogenetics, 10: 123-131.

Parkin, D.M., Whelan, S.L., Ferlay, J., Raymond, L.A. and Young, J. (1997). Cancer incidence in five continents. IARC Pub. 143, Lyon.

Roy, B. and Sikdar, N. (2003). Polymorphisms in drug metabolising genes and risk of head and neck squamous cell carcinoma. Int. J. Hum. Genet., 3(2): 99-108.

Sanders, B.G. and Kline, K. (1995). Nutrition immunology and cancer: an overview. Adv. Exp. Med. Biol., 369: 185-194.

Sankarnarayanan, R., Nair, M.K., Mathew, B., Balaram, P., Sebastian, P. and Dutt. S.C. (1992). Recent results in oral Cancer research in Kerala. Head and Neck, 14: 107-112.

Sarkar, D., Sharma, A. and Talukder, G. (1994). Chlorophyll and chlorophyllum as modifiers of genotoxic effects. Mut. Res. Reviews in Genetic Toxicology, 318: 239-247.

Sarkar, D., Sharma, A. and Talukder, G. (1996). Plant extracts as modulators of genotoxic effects. Bot. Rev., 62: 275-300.

Sen, S., Talukder, G. and Sharma, A. (1986). Carcinogenic, mutagenic and cytotoxic action of betel quid on mammalian systems. The Nucleus, 29(3): 169-182.

Sen, S., Talukder, G. and Sharma, A. (1989). Betel cytotoxicity: Review. J. Ethnopharmacol., 26: 217-247.

Sen, S., Agarwal, K., Mukherjee, A., and Sharma, A. (1994). Potentiation by caffeine on the frequencies of chromosome aberration. Drug. and Chem. Toxicol., 17: 113-124.

Sen Pathak, (2003). Cancer biology and genetic heterogeneity. Proc. Ind. Natl. Sci. Acad., 869: 11–29.

Sharma A and Talukder G (2000) Dietary intervention against environmental mutagenesis by plant extracts. J. Trp. Med. Plants, 1: 15-19.

Slaga, T.J., DiGiouanni, J., Winberg, L.D. and Budunova, I.V. (1995). Skin carcinogenesis: characteristics, mechanisms and prevention. Proc. Clin. Biol. Res., 391: 1-20.

Slaga, T.J., Budunova, I.V., Gimenez-Conti, I.B. and Aldaz, C.M. (1996). The mouse skin carcinogenesis model. J. Invest. Dermato. Symposium Proceedings, 1: 151-156.

Sreelekha, T.T., Ramadas, K., Pandey, M., Thomas, G., Nalinakumari, K.R. and Pillai, M.R. (2001). Genetic polymorphism of CYPlAl, GSTMl and GSTTl genes in Indian oral Cancer. Oral. Oncol., 37(7): 593-8.

Stitch, H.F., Mathew, B., Sankarnarayanan, R. and Nair, M.K. (1991a). Remission of oral precancerous lesions of tobacco/areca nut chewers following administration of beta carotene or vitamin A and maintenance of protective effect. Cancer Detect. Prev., 15(2): 93-98.

Stitch, H.F., Mathew, B., Sankarnarayanan, R. and Nair, M.K. (1991b). Remission of precancerous lesions in the oral cavity of tobacco chewers and maintenance of the protective effect of beta carotene or vitamin A. Am. J. Clin. Nutr., 53(Suppl): 298s-304s.

Wang, Z.Y., Agarwal, R., Bickers, D.R. and Mukhtar, H. (1991). Protection against ultraviolet B radiation-induced photocarcinogenesis in hairless mice by green tea polyphenols. Carcinogenesis, 12: 1527-1530.

Wang, Z.Y., Agarwal, R., Khan, W.A. and Mukhtar, H. (1992). Protection against benzo(a)pyrene and N-nitrosodiethylamine-induced lung and forestomach tumorigenesis in A/J mice by water extracts of green tea and licorice. Carcinogenesis, 13: 1491-1494.

Wang, Z.Y., Huang, M.T., Lou, Y.R., Xie, J.G., Reuhl, K.R., Newmark, H.L., Ho, C.T., Yang, C.S. and Conney, A.H. (1994). Inhibitory effects of black tea, green tea, decaffeinated black tea, and decaffeinated green tea on ultraviolet B light-induced skin carcinogenesis in 7,12- dimethylbenz(a)anthracene-initiated SK-l mice. Cancer Res, 54: 3428.

Warren, B.S., Naylar, M.F., Winberg, L.D., Yashimi, N., Volpe, I.P., Gimenez-conti, I. and Slaga, T.J. (1993). Induction and inhibition of tumor progression. Proc. Soc. Exp. Biol. Med., 202: 9-15.

Wattenberg, L. (1985). Chemoprevention of cancer. Cancer Res., 45: 1-8.

Weisburger, J.H., Hara, Y., Dolan, L., Luo, F.Q,. Pittman, B. and Zang, E. (1996). Tea polyphenols as inhibitors of mutagenicity of major classes of carcinogens. Mutat. Res., 371: 57-63.

Wu, P.F., Chiang, T.A., Wang, L.F., Chang, C.S. and Ko, V.C. (1998). Nitro-polycyclic aromatic hydrocarbon contents of fumes from heated cooking oils and prevention of mutagenicity by catechin. Mutat. Res., 403: 29-34.

Yamane, T., Hagiwara, N., Tateishi, M., Akachi, S., Rim, M., Okuzumi, J., Kitao, Y., Inagake, M., Kuwata, K. and Takahashi, T. (1991). Inhibition of azoxymethane-induced colon carcinogenesis in rat by green tea polyphenol fraction. Jpn. J. Cancer Res., 82: 1336-1339.

Yang, C.S. and Wang, Z.Y. (1993). Tea and cancer. J. Natl. Cancer. Inst., 85: 1038-1049.

Yang, C.S. (1997). Inhibition of carcinogenesis by tea. Nature, 389: 6647.

Yang, C.S., Lee, M.J., Chen, L.S. and Yang, G.Y. (1997). Polyphenols as inhibitors of carcinogenesis. Env. Hlth. Persp., 105: 971-976.

Yen, G.C. and Chen, H.Y. (1996). Relationship between antimutagenic activity and major components of various teas. Mutagenesis, 11: 37-41.

Yen, G.C., Chen, H.Y. and Peng, H.H. (1997). Antioxidant and pro-oxidant effects of various tea extracts. J. Agr. and Food Chem., 45(1): 30-34.

Yoshizawa, S., Horiuchi, T., Fujiki, H., Yoshida, T., Okuda, T. and Sugimura, T. (1987) Antitumor promoting activity of (–)-epigallocatechin gallate, the main constituent of "tannin" in green tea. Phytother Res, 1: 44-47.

Zavras, A.I., Wu, T., Laskaris, G., Wang, Y.F., Cartsos, V., Segas, J., Lefantzis, D., Joshipura, K., Douglass, C.W. and Diehl, S.R. (2002). Interaction between a SNP in the alcohol dehydrogen 3 gene alcohol consumption and oral cancer risk. Int. J. Cancer, 97(4): 526-530.

Zheng, W., Shu, X.O., Ji, B.T. and Gao, Y.T. (1996). Diet and other risk factors for cancer of the salivary glands: a population based case control study. Int. J. Cancer, 17: 194-198.

Ziegler, R.G. (1991). Vegetables, fruits and carotenoids and risk of cancer. Am. J. Clin. Nutr., 53 (Suppl.): 251-259.

Chapter

7

Cancer Prevention by Cancer Regression and Rejuvenation of Host's Defense System: Dual Role of Tea

Lakshmishri Lahiry, Deba Prasad Mandal,
Arindam Bhattacharyya, Gaurisankar Sa and Tanya Das
Bose Institute, Kolkata, India

ABSTRACT

Cancer is perhaps the most dreaded and destructive disease posing great threat to human beings world over. Despite significant advances in medical technology for diagnosis and treatment, selective killing of tumor cells without endangering the healthy normal tissues of the same tumor bearer still remains as complex as the disease itself. Many of the intrinsic defense systems of the host, like immune system, bio-transformation and aspects detoxification enzyme systems, DNA repair mechanisms, and cell cycle regulation and apoptosis, are adversely affected directly or indirectly, by the presence of the tumor, rendering the host incapable of mounting a successful anti-tumor therapy. Moreover, during most of the chemotherapeutic treatments, toxic and immunosuppressive effects further complicate an already weakened defense machinery of the tumor-bearing host. Recently, considerable attention has been paid in identifying naturally occurring chemopreventive substances capable of inhibiting, retarding or reversing the toxicity generated in the system due to toxic/carcinogenic insult. Tea, particularly its polyphenolic constituents, has emerged as a non-toxic, immunopotentiating, cancer preventing agent. In this review, we hypothesize that tea acts as a double-edged sword in the treatment of cancer by: (1) directly killing tumor cells via apoptosis and (2) protecting and potentiating the intrinsic defense machinery of the tumor-bearing host by virtue of its anti-inflammatory, anti-oxidative, anti-mutagenic effects and successful amelioration of tumor-induced immunosuppression.

INTRODUCTION

Cancer– an Overview

Cancer is one of the leading causes of death in industrialized nations. The mechanisms of carcinogenesis involve multiple stages of biochemical and molecular alterations in target cells. Diagnosis of and therapy for cancer are usually high-cost elements in health management. Thus, there has been increasing emphasis on research to develop an understanding of cancer, as well as the action of modulating factors, as the bases for prevention and therapy. The process of carcinogenesis involves the stepwise accumulation of genetic changes, ultimately leading to malignancy (DiGiovanni, 1992). There are three main steps:

(a) Initiation – this step in multistage carcinogenesis involves carcinogen induced genetic changes and appears to be rapid a well as irreversible. In this stage irreversible modification of DNA occurs perhaps resulting in one or more mutations;

(b) Promotion – this is a much slower step and is believed to involve selective and sustained hyperplasia, leading to the specific expansion of initiated cells into papillomas. These changes are believed to result from epigenetic mechanisms; and

(c) Progression – the process of tumor progression is characterized by a high level of genetic instability that leads to a number of chromosomal alterations.

To achieve full malignancy, cells must acquire certain transformation characteristics, including (a) self-sufficiency in growth signaling and limitless replicative potential; (b) unresponsive to antiproliferative signals, (c) evading apoptosis; (d) inducing and sustaining angiogenesis; and (e) acquiring the ability to invade and metastatize (Hanahan and Weinberg, 2002). This sequence of events presents many opportunities for intervention, with the aim of preventing, slowing down or reversing the transformation process.

As treatments for infectious diseases and the prevention of cardiovascular disease continue to improve, and due to the benefits of these treatments, the average life expectancy of man increases, cancer is likely to become the most common fatal disease in developed countries. Cancers are caused by the progressive growth of the progeny of a single transformed cell. Therefore, curing cancer requires that all the malignant cells be removed or destroyed without killing the patient. An attractive way to achieve this would be to use such an agent that would discriminate between the cells of the tumor and their normal cell counterparts and be selectively lethal to the malignant phenotypes. Since, cancer itself, as well as most of the drugs used for cancer therapy are toxic to the intrinsic defense machinery of the cancer bearer, successful cancer therapy has still remained an unfulfilled dream to scientists and medical professionals worldwide. In this regard, dietary phytochemicals, which are non-toxic to the normal cells of the tumor bearer, have gained immense importance.

CANCER AND THE INTRINSIC DEFENSE MACHINERY OF THE HOST: A TUG-OF-WAR

It is interesting to note that each organism has the built-in capacity to tolerate carcinogenic/toxic insult up to a certain limit, which of course depends on the physiological status of the individual and varies from one form of stress to another. Interestingly, interaction with the products of nature and those developed by human ingenuity does not normally lead to physiological catastrophe since human and other higher organisms have developed a number of defense mechanisms to protect themselves against toxic and carcinogenic insult. Among these, the most important are:

I. Immune system
II. Biotransformation and detoxication enzyme systems
III. Anti-oxidant system
IV. DNA repair mechanisms

However, during diseases like cancer or due to long term exposure to toxic drugs, these defense machineries of the host are jeopardized; consequently, the disease prevails, resulting in the patient's death. Given below is an overview of the detrimental effects of cancer on the host's defense systems.

(i) Cancer vis-à-vis Immune System Human beings and species of other higher organisms maintain intrinsic regulatory programs (host defense mechanism) to ensure protection from the carcinogenic attack of environmental cues and thereby help in prevention of cancer and many other diseases. However, continuous exposure of the body to various carcinogenic/toxic agents, developing cancer itself and intake of cancer drugs, may jeopardize the defense system of the host, ultimately leading to death. The immune system is responsible for early detection and destruction of newly transformed malignant cells. However, the genetic plasticity of cancer cells can lead to alterations that confer selective growth advantages to the tumor, some of which play a role in immune escape. There is ample evidence for the presence of tumor-associated antigens on a variety of tumors, although they are seemingly unable to elicit an adequate antitumor immune response (Tuting et al., 1997). Moreover, a number of mutations veiling tumor cells from host immune defenses have been well characterized but more recent studies suggest that a variety of tumors can also express products that are actually toxic for the immune effectors (Wojtowicz, 1997). A large number of cytokines, hormones, and other molecules secreted by tumors have been demonstrated to have immunomodulating properties, of them the most extensively studied immunosuppressive molecules secreted by tumors are transforming growth factor-beta, interleukin 10, and prostaglandin E2 (Wojtowicz, 1997). It has also been reported that the expression of Fas ligand on tumor cells has been implicated in their evasion of immune surveillance (Chen et al., 2003). Recent observations also implied apoptosis as the cause of tumor-induced immunosuppression (Sogn, 1998). However, there is hardly any study showing the tug-of-war between developing tumor and immune cells of the host and the underlying mechanism by which the developing tumor induces immune cell death.

(ii) Cancer and Detoxification Mechanisms of Host Disease and health status of the individual also influence detoxification activities. Since the majority of detoxification occurs in the liver, it is not surprising that impairment of normal liver function due to different diseases like cancer can lead to lower detoxification activity in general (Lee, 1995). The Pi (π) isoenzyme of Phase I enzymes is present in very low amounts in normal liver but its expression becomes very high with the onset of carcinogenesis (Jensson et al., 1985). As a consequence of Phase I biotransformation in detoxification, reactive molecules, which may be more toxic than the parent molecule and may cause damage to proteins, RNA, and DNA within the cell, are produced and are in turn further metabolized by Phase II conjugation. Therefore, any impairment in their activities due to cancer-induced suppression may make the situation even graver for the host (Ray et al., 2001).

(iii) Effect of Cancer on Host's Anti-oxidant System Due to cellular metabolism, reactive oxygen species (ROS) is generated within the cell. Accumulation of ROS oxidizes cellular fatty acids to lipid peroxy radicals that diffuse through membrane and interacts with DNA, resulting in a loss of cell homeostasis (Rice-Evans and Burdon, 1993). Antioxidant enzymes like catalase, superoxide dismutase, glutathione-S-transferase, peroxidase and many other enzymes prevent the oxidative damage within the cell (Zhang et al., 2001). It is now well recognized that with onset of cancer, there is a concurrent toxic manifestation in the form of oxidative stress within the host. Status of antioxidant enzymes, i.e., Superoxide dismutase (SOD), Catalase, glutathione-S-transferases (GST), peroxidases etc., which protect cells against oxidative damage, is almost invariably altered during carcinogenesis (Marx, 2001). Consequently, accumulated reactive oxygen species interacts with and modifies cellular proteins, lipid and DNA, which results in altered target cell function (Klaunig and Yong, 1998).

(iv) Cancer and DNA Repair Mechanism It is known that in a normal cell there is DNA repair mechanism to rectify any error which may otherwise lead to production of faulty proteins, and thereby result in defective cellular functions, and may even cause cancer. In fact, cancer occurs when mutagens interfere with the DNA repair mechanism (Phillips, 2001), induce mutation (Ames, 1998; Conilly, 1995) and disturb normal cell cycle progression machinery, thereby disrupting the balance between cell growth and cell death, leading to uncontrolled proliferation (Sell, 1993; Norbury and Hickson, 2001). On the other hand in cancer bearer, developing cancer inhibits the DNA repair mechanisms of normal cells of the host, thus disrupting their normal cell cycle regulation (Agarwal, 1992).

As is evident from the above discussions that cancer, once set in, causes severe immunodepression, depresses the detoxification and anti-oxidant systems and de-regulates DNA repair mechanisms (Fig. 7.1). Moreover, most

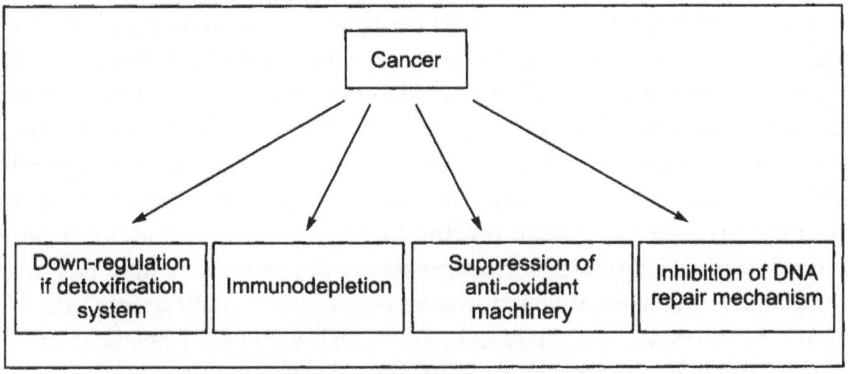

Fig. 7.1 Cancer-induced suppression of host's defense machinery.

of the cancer drugs suppress host's already compromised defense system – an account of which is being furnished below.

Why Cancer Therapy Fails

Mainstream medicine (relying upon surgery, chemotherapy and radiation) may initially appear successful in the treatment of cancer, but they have their own specific shortcomings. For the past 30 years, cancer therapies have experienced tremendous setbacks because of an associated toxic response, resulting in significant numbers of treatment-induced deaths rather than disease-induced fatalities. During the regimen of cancer chemotherapy, various popular and effective drugs in use these days exert concurrent toxic manifestations including oxidative stress, liver damage and immunosuppression in the tumor bearer (Gopalakrishna and Jaken, 2000). Many toxic metabolic products of the cancer drugs repress Phase I and II biotransformation and detoxication enzyme systems, which otherwise are required to detoxify and eliminate the unwanted metabolic products from the body. As a result, hepatotoxic marker enzymes like serum glutamate oxaloacetate transaminase (SGOT), serum glutamate pyruvate transaminase (SGPT) and serum alkaline phosphatase (ALP) activities are often elevated (Bhimani, 1993). During most of the chemotherapeutic treatments, immunosuppressive effects are almost always observed. Beyond a level, each chemical can cause immunosuppression (Ray and Das, 1998) thereby leaving the host susceptible to any kind of carcinogenic/toxic attack (Katiyar and Mukhtar 2001; Van Loveren et al., 1995). Bone marrow suppression is frequently associated with such a therapy (Sosin and Handa, 2003). Moreover, most of the cancer drugs have been reported to cause simultaneous apoptosis in splenocytes of the tumor bearer (Ferraro et al., 2000). Even nurses in cancer treatment settings, who are exposed continuously to toxic chemotherapy drugs such as fluoruacil, cyclophosphamide and others, have been found to have more abnormalities in lymphocyte chromosome structures. All this information indicates that cancer itself, as well as the existing cancer drugs, besides regressing tumor, suppress the defense systems of the host, thereby making the situation even worse. Thus, from the point of view of successful cancer therapy, selective targeting and low toxicity for normal host tissues are fundamental requisites for proposed chemopreventive agents since a non-selective mechanism of action may lead to potentially severe immune toxicity (Berlin et al., 1997; Bergan et al., 1999; Ewer et al., 1999; Eckhardt et al., 1999; Gibbs, 2000; Standler et al., 2000; Steele, 2002; Stremmel et al., 2002). Scientists all over the world are, therefore, in search of such an agent that may reconstitute the intrinsic defense system of the host to ultimately prevent/cure cancer.

DIETARY THERAPY

Recently, considerable attention has been focused on identifying naturally occurring chemopreventive substances capable of inhibiting, retarding, or reversing the process of multistage carcinogenesis. Wide arrays of phenolic substances, particularly flavonoids, which are present in dietary and medicinal plants, have been reported to possess substantial anti-carcinogenic and anti-mutagenic effects (Lin and Liang, 2000). These flavonoids are naturally occurring low molecular weight polyphenolic compounds widely distributed in fruits, vegetables and beverages. People who eat diets rich in fruits and vegetables have lower incidences of diseases such as cancer. Numerous experimental studies have examined the role of specific flavonoids in disease prevention. For example, increased flavonoid intake was associated with decreased risk of carcinogenesis (Hertog et al., 1993). The majority of these naturally occurring phenolic compounds retain antioxidative and anti-inflammatory properties, which appear to contribute to their chemopreventive activity (Hertog et al., 1993). Because of their safety and the fact that they are not perceived as medicine, food derived products have high potential for development as chemopreventive and therapeutic agents that may find widespread and long-term use. Moreover *in-vitro* studies are now being conducted to identify the molecular targets within cancer cells that are modulated by these dietary constituents (Manson, 2003). Tea is one of such dietary substances having diverse biomodulatory activities and therefore, may be a good candidate as a safe and potential anticancer agent.

WHY TEA?

The inhibition of carcinogenesis by tea has been demonstrated in cancers of skin, lung, esophagus, stomach, liver, duodenum and small intestine, pancreas, colon, bladder, prostate and mammary glands (Dreosti et al., 1997; Katiyar and Mukhtar, 1997; Yang et al., 1998). Green tea has been shown to possess cancer chemopreventive effects in a wide range of target tumors. Furthermore, the Japanese and Chinese populations who regularly consume green tea, have one of the lowest incidence of prostate cancer (Gupta et al., 1999). Multiple biological effects of flavonoids have been described, among them are anti-inflammatory, anti-allergic, anti-hemorrhagic, antimutagenic, anti-neoplastic and hepatopreventive activities. The flavonoids in green tea have been studied intensely in many laboratories. The major fraction of these flavonoids is the group of catechins in green tea. The chemopreventive effects of green tea against tumorigenesis and tumor growth have been attributed to the biochemical and pharmacological activities of its polyphenolic constituents, most notably Epigallocatechin gallate (EGCG), present therein (Gupta et al., 2000). Epidemiological studies also suggest a protective effect of green tea consumption on some cancer types in humans (Bushman, 1998). Black tea has also been shown to inhibit tumorigenesis in animal model systems, including skin, lung, colon, esophagus, pancreas, liver and mammary glands (Lu et al., 2000) and to block neoplastic transformation in mouse epidermal cell line (Liang et al., 1999). However, besides some scattered reports and supporting epidemiological evidences on the protective role of black tea, the detailed molecular mechanisms underlying its antitumor effect are still unclear and inconclusive, although the worldwide production and consumption of black tea far exceeds that of green tea (Bickers and Athar, 2000). This review describes the selective anti-carcinogenic effect of tea without harming the normal cells of the tumor bearer as well as the molecular mechanisms that may be involved.

TEA AND ITS CONSTITUENTS – A HISTORIC VIEW

From its legendary discovery in the year 2737 BC by the Chinese emperor (a scholar/herbalist) Sheng Nung when a leaf accidentally fell into a pot of boiling water, to the universally enjoyed beverage it is today, tea has played a significant role in human history. It has many health friendly properties in addition to its delicious taste as a drink. Moreover, various reports from different parts of world have demonstrated its anti-tumor, anti-cancer and immunomodulatory properties.

Tea has been used as a daily beverage and crude medicine in China and India for several thousand years. This beverage is primarily manufactured as green, black or oolong tea according to the degree of fermentation involved. Most Japanese and Chinese people in northern China prefer green tea, whereas Indians, Americans and Europeans prefer black tea. Oolong and paochong tea are favored in Taiwan and parts of China. The composition of tea varies with the species, season, age of the leaf (plucking position), climate, and horticultural practices. Non-fermented green tea contains several groups of polyphenols that account for up to 40% of the dry leaf weight. Flavanols are the major polyphenolic fraction of green tea and include catechins (Yang et al., 2000), e.g., (–)-epigallocatechin-3-gallate

(EGCG), (–)-epigallocatechin (EGC), (–)-epicatechin-3-gallate (ECG), (–)-epicatechin (EC), (+)-gallocatechin (GC) and catechin (C) (Fig. 7.2) (Graham, 1992). In the manufacture of black tea, the "fermentation" process causes green tea catechins to oxidize and form oligomeric flavanols, including theaflavins, thearubigin and other oligomers. Theaflavins are a mixture of theaflavin (TF-1), theaflavin- 3-gallate (TF-2a), theaflavin-3′-gallate (TF-2b) and theaflavin-3,3′-digallate (TF-3) (Fig. 7.2). However, thearubigins are the most abundant phenolic fraction in black tea. They vary greatly in molecular weight, and have a structure that is not yet well characterized. Partially fermented oolong or paochong tea contains both green tea catechins, black tea theaflavins and, possibly thearubigins. Some components in oolong or paochong tea, such as, proanthocyanidins, are less well characterized and they may be important in disease prevention.

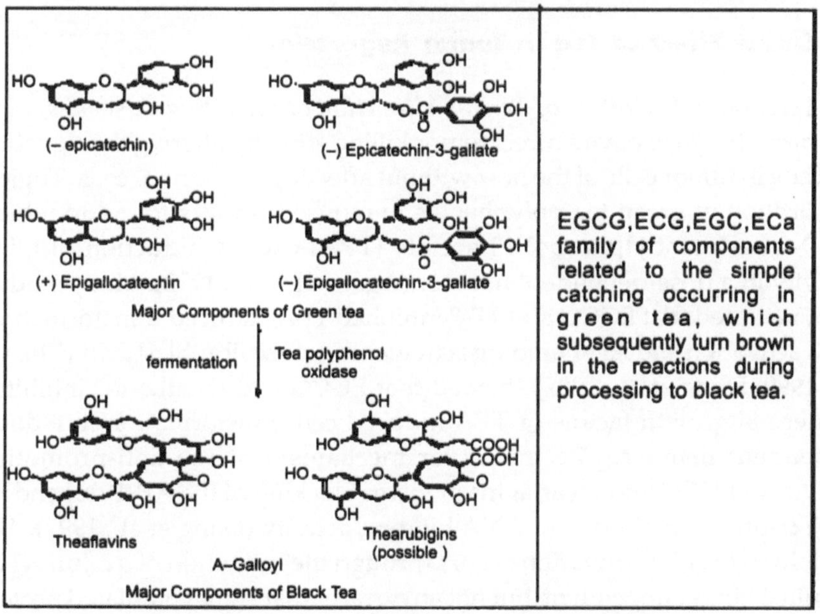

Fig. 7.2 Major components of green tea and black tea

HYPOTHESIS

We hypothesize that tea may act as an effective anti-cancer agent by: (1) directly killing tumor cells via apoptosis, and (2) protecting and potentiating the intrinsic defense machinery of the tumor-bearing host

(Fig. 7.3). Some studies that are currently being carried out in our laboratory and others supporting this hypothesis are detailed below.

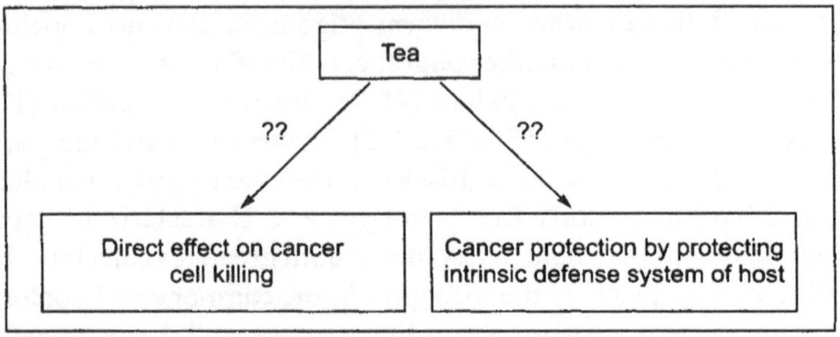

Fig. 7.3 Can tea regress tumor by (i) killing cancer cells directly and (ii) protecting and potentiating the intrinsic defense machinery of the tumor-bearing host?

(1) Direct Effect of Tea in Tumor Regression

(a) ***Anticancer Activities of Tea by Alteration of Signaling Cascades*** Tea has been shown to have a direct tumor killing effect by altering the signaling cascade in tumor cells of the host without affecting the normal cells. Topical application of green tea polyphenols to mouse skin was found to inhibit 12-O-tetradecanoylphorbol-13-acetate (TPA)-caused induction of ODC activity in a dose-dependent manner (Agarwal et al., 1992). Many studies demonstrated that EGCG and TF-3 inhibited TPA-induced transformation, PKC activation and AP-1 binding activities in mouse fibroblast cells (Chen et al., 1999; Dong et al., 1997) showed that EGCG and theaflavins inhibited epidermal growth factor- or TPA-induced cell transformation in a dose-dependent manner. The molecular mechanism of the anti-promotion activity of EGCG and theaflavins involves blocking of the AP-1-dependent transcriptional activity and DNA binding activity (Dong et al., 1997). The inhibition of AP-1 activation occurs through the inhibition of a c-Jun NH2-terminal kinase-dependent, but not an extracellular signal regulated protein kinase ErK 1-dependent or Erk2- dependent pathway.

Several reports demonstrated that the major and most potent component of tea polyphenols, EGCG and TF-3, strongly inhibited the kinase activity of EGF and PDGF receptors as induced by EGF or PDGF, respectively (Liang et al., 1997, Liang et al., 1999). Additional studies showed that both EGCG and TF-3 blocked EGF from binding to its receptor (Lea et al., 1993). These results suggested that EGCG and TF-3 have an anti-proliferation effect on tumor cells, and that the molecular mechanisms of this effect may block the growth factor binding to its receptor, and subsequently through inhibition of receptor kinase activity, suppress mitogenic signal transduction. These effects might have an inhibitory role on proliferation and cell cycle progression, even on apoptosis. It has been reported that EGCG or green and black tea polyphenols strongly inhibited DNA synthesis in DS19 mouse erythroleukemia cells and HTC rat hepatoma cells (Chen et al., 1998). The effect of theaflavins on the growth of SV-40 virally transformed WI38 human fibroblasts (WI38VA) were compared with that of normal WI38 cells (Chen et al., 1998). The results indicated that tea polyphenols had a pronounced growth inhibitory effect on cancerous cells (WI38VA) but not on their normal counterparts (Lu et al., 2000). Similar differences in growth inhibition were also observed between a human colorectal cancer cell line (Caco-2), a breast cancer cell line (Hs578T) and their respective normal counterparts. Moreover, it was revealed that tea polyphenols displayed strong growth of inhibitory effects against a lung tumor cell line (Yang et al., 1998). The results suggested that tea polyphenols-induced production of H_2O_2 may mediate apoptosis, and that this may contribute to the growth inhibitory activities of tea polyphenols *in vitro*.

(b) Tea-induced Cell Cycle Regulation and Induction of Apoptosis Aberrant proliferation and modulated apoptosis leading to impaired cellular homeostasis represent crucial early events in the multi-step carcinogenic process. Effect of tea on cell cycle has been well documented. It has been reported that tea polyphenols inhibited the growth of human lung cancer cell line PC-9, with G2/M arrest (Fujiki et al., 1998). EGCG can inhibit cervical cancer cell growth in human papilloma virus HPV-16 associated cervical cancer cell line, CaSki cells, through the induction of apoptosis and cell cycle arrest as well as regulation of gene expression *in vitro* (Ahn et al., 2003). Cells were arrested at the G1 phase suggesting that cell cycle arrests might precede apoptosis in these cell lines upon EGCG treatment (Ahn et al., 2003). The antitumor effects of the green tea compound epigallocatechin-3-gallate (EGCG) have been studied in detail in head and neck squamous cell carcinoma (HNSCC) cells (Masuda et al., 2001). Treatment with EGCG increased the proportion of cells in the G(1) phase of the cell cycle and induced apoptosis. In cells treated with EGCG, there was a decrease in the cyclin D1 protein, an increase in the p21 (Cip1) and p27 (Kip1) proteins, and a reduction in the hyperphosphorylated form of pRB, changes that may account for the arrest in G1. EGCG also caused a decrease in the Bcl-2 and Bcl-XL proteins, an increase in the Bax protein, and activation of caspase 9, suggesting that EGCG induces apoptosis via a mitochondrial pathway

(Masuda et al., 2001). In one study, it was shown that treatment of A431 (Human epidermoid carcinoma) cells with green tea polyphenols and its components, epigallocatechin-3-gallate, epigallocatechin, and epicatechin-3-gallate, resulted in the formation of internucleosomal DNA fragments, characteristic of apoptosis (Ahmad et al., 1997). Treatment with epigallocatechin-3-gallate also resulted in apoptosis in human keratinocyte carcinoma, human prostate carcinoma cells and mouse lymphoma cells (Ahmad et al., 1997). The DNA cell cycle analysis showed that in A431 cells, epigallocatechin-3-gallate treatment resulted in arrest in the G0-G1 phase of the cell cycle and a dose-dependent apoptosis (Ahmad et al., 1997). It is now well recognized that whether a cell becomes committed to apoptosis partly depends upon the balance between proteins that mediate cell death, e.g., p53, p21, Bax, and proteins that promote cell viability, e.g., Bcl-2 (Oltvai et al., 1993). A report from our laboratory has already established the relationship between p53 status, p21 induction, Bcl-2/Bax ratio, cell cycle deregulation and apoptosis in black tea-treated tumor cells (Bhattacharyya et al., 2003 a). The results imply an apoptosis enhancing capability of black tea in Ehrlich's ascites carcinoma (EAC) cells that was effected by modulating the tumor cell cycle as well as the balance between pro- and anti-apoptotic factors.

(2) Indirect Effect of Tea in Cancer Prevention: Tea-induced Tumor Regression by Rejuvenation of Host's Intrinsic Defense Machineries

(a) ***Tumor Regression by Amelioration of Tumor-induced Immunosuppression*** The tumor microenvironment influences the functional potential of immune cells. Escape from immune surveillance prefigures the rapid progression of cancers (Chouaib et al., 1997). Various immune escape mechanisms in cancer have been proposed (Pawelec et al., 1997). Certain cancer cells may secrete immunosuppressive factors to modify the host immune responses (Zvetkova et al., 2001; Bhattacharyya et al., 2004). In addition, anaemia is a common complication of malignancies. Cancer-related anaemia may occur as a direct effect of the neoplasm, it may be due to products of the cancer, or it may develop as a result of the cancer treatment itself. Tea has been found to have a protective function on immune system in cancer (Zvetkova et al., 2001) although the underlying mechanism is still unclear. We have already analysed that anticancer dose of black tea significantly protects immune cells from tumor-induced depletion (Bhattacharyya et al., 2004). All this information leads us to conclude that by potentiating the host's immune system, tea helps in regressing tumor.

(b) Antiinflammatory Effect of Tea Plays a Role in Cancer Prevention The induction of inflammation in skin mediated by TPA is believed to be governed by cyclooxygenase (COX), lipooxygenase and ODC. These markers of inflammatory responses are important for skin tumor promotion. It was demonstrated that pre-application of black tea polyphenols to that of TPA resulted in significant inhibition of TPA-caused induction of epidermal ODC and of COX enzyme activities (Katiyar and Mukhtar, 1997). Nitric oxide (NO) radical has a wide biological role in modulating physiological and pathophysiological processes (Moncada, 1992). During infection and inflammatory responses, the formation of NO is increased, which might promote carcinogenesis. Various studies showed both EGCG and TF-3 inhibited the induction of NO in thioglycollate-elicited and lipopolysaccharide (LPS)-activated peritoneal or RAW264.7 macrophages (Lin and Lin, 1997). In fact, tea polyphenols decreased the activity and protein levels of inducible nitric oxide synthase (iNOS) by reducing the expression of iNOS mRNA, and that the reduction could occur as a result of prevention of binding of the nuclear factor-B to the iNOS promoter, thereby inhibiting the induction of iNOS transcription (Lin and Lin, 1997).

(c) Antioxidative Effects of Tea in Prevention of Cancer Development It is known that oxidative DNA damage is one of the most important reasons behind cancer development. Tea and its polyphenols have been found to possess anti-oxidative properties and thereby possibly prevent a normal cell from becoming malignant. The antioxidant EGCG sets in motion a process called apoptosis, resulting in the death of cancer affected cells only, and not the healthy ones. EGCG reinforces the body's natural antioxidant system and promotes elimination of damaging oxygen molecules. It has been reported that tea polyphenols are strong scavengers of superoxide, hydrogen peroxide, hydroxy and other free radicals produced by various chemicals (Najo et al., 1999). The antioxidant properties of various tea polyphenols have been extensively investigated (Chen and Ho, 1994). Their study showed that the 1,1-diphenyl-2-picrylhydrazyl (DPPH) radical-scavenging ability of various black tea polyphenols was TF-2 > TF-1 > TF. The DPPH radical- scavenging activity was proportional to the number of –OH groups in the catechins or theaflavins. However theaflavins showed less lipid oxidation–inhibition activity than EGCG. But, all the theaflavins exhibited the same ability to inhibit the production of superoxide. Oxygen–radical absorbance capacity assay showed that EGCG has maximum scavenging effect on peroxy radical (Lin et al., 1996). It has been demonstrated that green tea polyphenols inhibited both cigarette smoke- and H_2O_2-induced DNA breakage in cultured lung cells (Leanderson et al., 1997). Green tea, black tea and EGCG were shown to block the production of oxygen free radicals derived from 2-amino-3-methylimidazo[4,5-f] quinoline (IQ) in the presence of NADPH-cytochrome P450 reductase (Hasaniya et al., 1997). Catechin polyphenols could also decrease the peroxynitrite- induced nitration of tyrosine and protect apolipoprotein B-100 of LDL from peroxynitrite induced modification of critical amino acids, which contribute to its surface charge (Pannala et al., 1997).

Recently, it has been observed from several studies that black tea and green tea offered protection against oxidative damage to red blood cells induced by a variety of inducers, e.g., H_2O_2, primaquine, 2,2'-azo-bis(2-amidinopropane) dihydrochloride (AAPH), phenylhydrazine (PHx), Cu2+-ascorbic acid, and the xanthine/xanthine oxidase system (Grinberg et al., 1997; Zhang et al., 1997; Halder and Bhaduri 1998). Oral administration of green or black tea leaf powder inhibited the lipid peroxidation of liver induced by tert-butyl hydroperoxide in rats (Sano et al., 1995) while in the kidney, the antioxidant effect was observed only for the green tea-fed group. The antioxidant effect of green and black tea on plasma antioxidant potential in man has been demonstrated (Serafini et al., 1996). Their results showed that ingestion of tea produced a significant increase of human plasma antioxidant capacity. It has been shown that (+)-catechin prevented human plasma from undergoing oxidation induced by [2,2'-azobis-(2-amidinopropane) clorhidrate] (AAPH) or [2,2'-azobis-(2,4-valeronitrile)] (AMVN) (Lotito and Fraga, 1998).

In addition, several reports indicated that tea intake might enhance the levels of antioxidant defense enzymes, such as SOD and catalase. The results indicated that increased activities of catalase and glutathione peroxidase were observed in the hepatic, pulmonary and small bowel. It has been found that treatment of rats with tea aqueous extracts as the sole drinking fluid for 4 weeks, at concentrations consumed by humans, could increase the UDP-glucuronosyl transferase activity (Bu-Abbas et al., 1995). The structure of tea polyphenols might consist of strong metal ion chelators, such as iron and copper, which are required for generation of reactive oxygen radicals by means of Fenton and Haber-Weiss reactions. Moreover, it has been suggested that the antioxidant properties of theaflavins and their gallate ester might work through chelating iron and copper (Miller et al., 1996). In our laboratory, we also observed that black tea protects the antioxidant system of the tumor-bearing host.

(d) ***Modulation of Metabolizing/Detoxifying Enzymes by Tea*** In the case of cancer chemotherapy, various popular and effective drugs exert concurrent toxic manifestations including oxidative stress and liver damage in the tumor bearer. Most procarcinogens require metabolic activation by metabolite enzymes, such as phase I and II enzymes, in order to convert to electrophiles before they can exert carcinogenic effects (Conney et al., 1982). As a result, glutamate oxaloacetate transaminase (GOT), glutamate pyruvate transaminase (GPT) and alkaline phosphatase are elevated in serum (Wolff et al., 1998). Tea has also been shown to effectively inhibit tumorigenesis induced by various carcinogens in the skin, lung, forestomach, pancreas, liver and esophagus of rodents by activating the detoxification systems of the host (Conney et al., 1982). Our findings indicate that anti-cancer dose of this popular beverage can ameliorate cancer-induced liver toxicity and hemotoxicity, thereby indicating that it can be a good candidate as a safe and potential anti-cancer agent. Activation of host detoxification system by black tea has also been confirmed by our work (Bhattacharyya et al., 2003 b). Green tea polyphenols have been found to inhibit rat liver microsomal mono-oxygenase activities, including Phase I enzymes of aryl hydrocarbon hydroxylase, 7-ethoxyresorufin O-deethylase, and 7-ethoxycoumarin O-deethylase (Wang et al., 1998). Another report indicated that treatment of rats with green and black tea for 4 or 6 weeks caused significant induction of cytochrome P450 enzymes, such as CYP1A2, CYP1A1, CYP2B, and CYP4A1 (Sohn et al., 1994; Bu-Abbas et al., 1994). Induced Phase I enzymes might be metabolic activators of carcinogens, e.g., CYP1A2 may play an important role in the activation of aromatic and heterocyclic amine procarcinogens. However, it has been reported that decaffeinated tea polyphenols did not cause significant induction of CYP enzymes. It is possible that the induction of CYP1A2 was an effect of caffeine (Chen et al., 1996).

(e) *Inhibition of Mutagenicity* Black and decaffeinated black tea were found to significantly inhibit the mutagenicity induced by three indirect-acting dietary carcinogens, Glu-p-1, benzo(a) pyrene (B[a]P) and nitrosopyrrolidine, in a bacterial test system (Bu-Abbas et al., 1996). Tea polyphenols sharply decreased the mutagenicity of a number of aryl- and heterocyclic amines, of aflatoxin B1, B[a]P, 1, 2-dibromoethane, and more selectively, of 2-nitropropane in *Salmonella typhimurium* (Weisburger et al., 1996). The relationship between the chemical content and antimutagenic activity of various tea extracts has been studied (Yen and Chen, 1996). The amount of catechins in various tea extracts was in the order: green tea > oolong tea > paochong tea > black tea. The tea extracts of oolong and paochong teas markedly inhibited the mutagenicity of various carcinogens. Recently, various groups examined the antimutagenic properties of several black tea extracts against distinct groups of environmental mutagens. The results suggested that different degrees of tea fermentation might produce distinct types of antimutagenic compounds effective against their corresponding mutagens (Hour et al., 1999). Fumes from cooking oils were found to have mutagenic activity. The results obtained indicated that fumes from cooking oils contained nitro- and polycyclic aromatic hydrocarbons (PAHs) which may be a risk factor for lung cancer among cooks and that the amounts of carcinogens could be reduced by adding catechin (Wu et al., 1998). Black tea polyphenols were more potent inhibitors of mutagenicity than green tea ones caused by the food mutagen 2-amino-1-methyl-6-phenylimidazo[4,5-b] pyridine (PhIP) in the Salmonella test system (Apostolides et al., 1996). Moreover, tea extracts have been shown to inhibit 2-amino-3-methylimidazo [4,5-f] quinoline (IQ) and B[a]P-induced mutagenicity in a Salmonella/ microsome assay (Chen and Yen, 1997). Their study revealed that tea extracts could inhibit the cytochrome P450-mediated metabolism of IQ and B[a]P into their ultimate mutagenic metabolite form, and interact with both promutagens and their metabolites in such a way as to reduce their mutagenic potentials. Another study also suggested that catechin might protect against such diverse reactive intermediates as free radicals and electrophiles forms during the metabolic activation of IQ (Hernaez et al., 1997).

CONCLUSION

Based on the results of numerous laboratory studies, including ours, it can be reiterated that tea consumption might reduce the risk of certain cancers. This review leads us to conclude that tea, the popular beverage, may directly act as an anti-cancer agent by killing tumor cells, or it may act as a "rescue drink" that strengthens the defense mechanism of the host, which otherwise may get suppressed due to developing cancer (Fig. 7.4). The knowledge of tea and its polyphenols adds a new dimension to our understanding of the use of dietary constituents either during therapy of cancer patients or as a preventive measure in high-risk individuals who work in a polluted environment containing carcinogenic/toxic chemicals.

ACKNOWLEDGEMENT

Work done in authors' laboratory has been sponsored by NTRF, Kolkata, and CSIR, New Delhi.

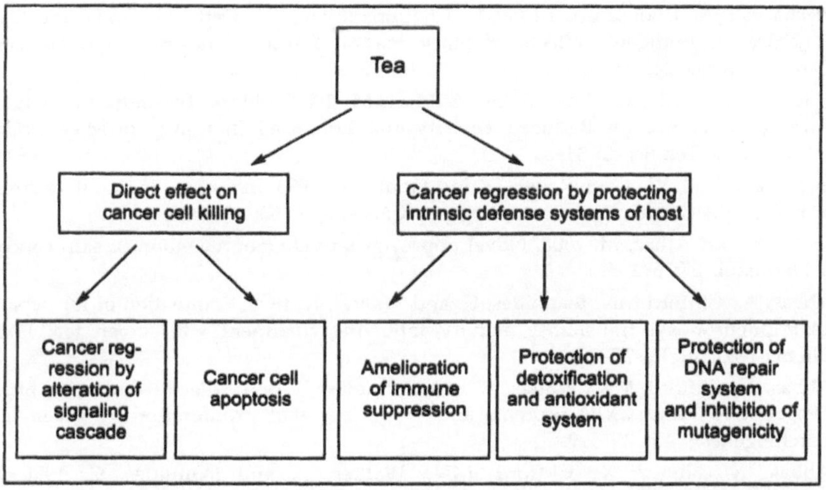

Fig. 7.4 Tea-induced cancer prevention by (i) directly killing cancer cells and (ii) rejuvenating host's suppressed defense machinery.

REFERENCES

Agarwal, R., Katiyar, SK., Zaidi, S.I. and Mukhtar, H. 1992. Inhibition of skin tumor promoter-caused induction of epidermal ornithine decarboxylase in SENCAR mice by polyphenolic fraction isolated from green tea and its individual epicatechin derivatives. Cancer Res. 52: 3582-8.

Ahmad, N., Feyes, D.K., Nieminen, A.L., Agarwal, R. and Mukhtar, H. 1997. Green tea constituent epigallocatechin-3-gallate and induction of apoptosis and cell cycle arrest in human carcinoma cells. J. Natl. Cancer Inst. 89: 1881–6.

Ahn, WS., Huh, SW., Bae, SM., Lee, IP., Lee, JM., Namkoong, SE., Kim, CK. and Sin JI. 2003. A major constituent of green tea, EGCG, inhibits the growth of a human cervical cancer cell line, CaSki cells, through apoptosis, G(l) arrest, and regulation of gene expression. DNA Cell Biol. 22: 217–24.

Ames, BN. 1989. Mutagenesis and carcinogenesis: endogenous and exogenous factors. Environ. Mol. Mutagen. 16: 66–77.

Apostolides, Z., Balentine, DA., Harbowy, ME. and Weisburger, JH. 1996. Inhibition of 2-amino-1-methyl-6-phenylimidazo[4,5-b]pyridine (PhIP) mutagenicity by black and green tea extracts and polyphenols. Mutat. Res. 359: 159–63.

Bergan, RC., Reed, E., Myers, CE., Headlee, D., Brawley, O., Cho, HK., Figg, WD., Tompkins, A., Linehan, WM., Kohler, D., Steinberg, SM. and Blagosklonny, MV. 1999. A Phase II study of high-dose tamoxifen in patients with hormone-refractory prostate cancer. Clin. Cancer Res. 5: 2366–73.

Berlin, J., Tutsch, KD., Hutson, P., Cleary, J. Rago, P., Arzoomanian, RZ., Alberti, D., Feierabend, C. and Wilding, G. 1997. Phase I clinical and pharmacokinetic study of oral carboxyamidotriazole, a signal transduction inhibitor. J. Clin. Oncol. 15: 781–89.

Bhattacharyya A, Mandal D, Lahiry L, Sa G, and Das T. 2004. Black Tea Protects Immunocytes from Tumor-induced Apoptosis by Changing Bcl-2/Bax ratio. Cancer Letter. 209: 147–54.

Bhattacharyya, A., Choudhuri, T., Pal, S., Chattopadhyay, SK., Datta, G., Sa, G. and Das, T. 2003a. Apoptogenic effects of black tea on Ehrlich's ascites carcinoma cell. Carcinogenesis 24: 75–80.

Bhattacharyya, A., Sa, G., Das, T. and Siddiqi, M. 2003b. Black Tea-Induced Cellular Survival: Evidence for Reduced Toxicity and Enhanced Immunity in Mice Under Stress. Int. J. Tea Sci. 2: 34–39

Bhimani, RS., Troll, W., Grunberger, D. and Frenkel K. 1993. Inhibition of oxidative stress in HeLa cells by chemopreventive agents. Cancer Res. 53: 4528–33.

Bickers, DR. and Athar, M. 2000. Novel approaches to chemoprevention of skin cancer. J. Dermatol. 27: 691–5.

Bu-Abbas, A., Clifford MN, Ioannides, C. and Walker, R. 1995. Stimulation of rat hepatic UDP-glucuronosyl transferase activity following treatment with green tea. Food Chem. Toxicol. 33: 27–30.

Bu-Abbas, A., Clifford, MN., Walker, R. and Ioannides, C. 1994. Selective induction of rat hepatic CYP1 and CYP4 proteins and of peroxisomal proliferation by green tea. Carcinogenesis 15: 2575–9.

Bu-Abbas, A., Nunez, X., Clifford, MN., Walker, R. and Ioannides, C. 1996. A comparison of the antimutagenic potential of green, black and decaffeinated teas: contribution of flavanols to the antimutagenic effect. Mutagenesis 11: 597–603.

Bushman, JL. 1998. Green tea and cancer in humans: a review of the literature. Nutr. Cancer. 31: 15.1-9.

Clen, CW. and Ho, CT. 1994. Antioxident properties of polyphenols extracted from green and black teas. J. Food Lipids 2: 35–46.

Chen, HY. and Yen, GC. 1997. Possible mechanisms of antimutagens by various teas as judged by their effects on mutagenesis by 2-amino-3-methylimidazo[4,5-f]quinoline and benzo[a]pyrene. Mutat. Res. 393: 115–22.

Chen., L., Bondoc, FY., Lee MJ, Hussin AH, Thomas, PE. and Yang, CS. 1996. Caffeine induces cytochrome P4501A2: induction of CYP1A2 by tea in rats. Drug Metab. Dispos. 24: 529–33.

Chen, YC., Liang, YC., Lin-Shiau, SY., Ho, CT. and Lih, JK. 1999. Inhibition of TPA-induced protein kinase C and transcription activator protein-1 binding activities by theaflavin-3,3'-digallate from black tea in NIH3T3 cells. J. Agric. Food Chem. 47: 1416–21.

Chen, YL., Chen, SH. and Wang, JY., and Yang, BC. 2003. Fas ligand on tumor cells mediates inactivation of neutrophils. J. Immunol. 1711: 1183–91.

Chen, ZP., Schell, JB., Ho, CT. and Chen, KY. 1998. Green tea epigallocatechin gallate shows a pronounced growth inhibitory effect on cancerous cells but not on their normal counterparts. Cancer Lett. 129: 173–9.

Chouaib, S., Asselin-Paturel, C., Mami-Chouaib, F., Caignard, A. and Blay, JY. 1997. The host-tumor immune conflict: from immunosuppression to resistance and destruction. Immunol Today 18: 493–497.

Conilly, RB. 1995. Cancer and non-cancer risk assessment: not so different if you consider mechanisms. Toxicology 102: 179–88.

Conney, AH. 1982. Induction of microsomal enzymes by foreign chemicals and carcinogenesis by polycyclic aromatic hydrocarbons: G.H.A. Clowes Memorial Lecture. Cancer Res. 42: 4875–917.

DiGiovanni, J. 1992. Multistage carcinogenesis in mouse skin. Pharmacol Ther. 54: 63–128.

Dong, Z., Ma, W., Huang, C. and Yang, CS. 1997 Inhibition of tumor promoter-induced activator protein 1 activation and cell transformation by tea polyphenols, (–)-epigallocatechin gallate, and theaflavins. Cancer Res. 57: 4414–9.

Dreosti, IE., Wargovich, MJ, and Yang CS. 1997. Inhibition of carcinogenesis by tea: the evidence from experimental studies. Crit. Rev. Food Sci. Nutr. 37: 761–70.

Eckhardt, SG., Rizzo, J., Sweeney, KR., Cropp, G., Baker, SD., Kraynak, M., Kuhn, JG., Vitalona-Calero, MA., Hammond, L., Weiss, G., Thurman, A., Smith, L., Drengler, R., Eckardt, JR., Moczygemba, J., Hannah, AL., Von Hoff, DD. and Rowinsky, EK.1999. Phase I and pharmacologic study of the tyrosine kinase inhibitor SU101 in patients with advanced solid tumors. J. Clin. Oncol. 17: 1095–104.

Ewer, MS., Gibbs, HR., Swafford, J. and Benjamin, R.J. 1999. Cardiotoxicity in patients receiving transtuzumab (Herceptin): primary toxicity, synergistic or sequential stress, or surveillance artifact? Semin. Oncol. 26: 96–101.

Feraro, C., Quemeneurl, A. and Prigent, AF. 2000. Anthracyclines trigger apoptosis of both G0-G1 and cycling peripheral blood lymphocytes and induce massive deletion of mature T and B cells. Cancer Res. 60: 1901–07.

Fujiki, H., Suganuma, M., Okabe, S., Sueoka, N., Komori, A., Sueoka, E., Kozu, T., Tada, Y., Suga, K, Imai, K. and Nakachi, K. 1998. Cancer inhibition by green tea. Mutat Res. 402: 307–10.

Gibbs, JB. 2000. Mechanism-based target identification and drug discovery in cancer research. Science 287: 1969–73.

Gopalakrishna, R. and Jaken S. 2000. Protein kinase C signaling and oxidative stress. Free Radic. Biol. Med. 128: 1349–61.

Graham, HN. 1992. Green tea composition, consumption, and polyphenol chemistry. Prev. Med. 21: 334-50.

Grinberg, LN., Newmark, H., Kitrossky, N., Rahamim, E., Chevion, M. and Rachmilewitz, EA. 1997. Protective effects of tea polyphenols against oxidative damage to red blood cells. Biochem. Pharmacol. 54: 973–8.

Gupta, S, Ahmad N. and Mukhtar, H. 1999. Prostate cancer chemoprevention by green tea. Semin. Urol. Oncol. 17: 70–6.

Gupta S., Ahmad, N., Nieminen, AL. and Mukhtar, H, 2000. Growth inhibition, cell-cycle dysregulation, and induction of apoptosis by green tea constituent (–)-epigallocatechin-3-gallate in androgen-sensitive and androgen-insensitive human prostate carcinoma cells. Toxicol. Appl. Pharmacol. 164: 82–90.

Halder, J. and Bhaduri, AN. 1998. Protective role of black tea against oxidative damage of human red blood cells. Biochem. Biophys. Res. Commun. 244: 903–7.

Hanahan, D. and Weinberg, R. 2002. The hallmarks of cancer. Cell. 100: 57–70.

Hasaniya, N., Youn, K., Xu, M., Hemaez, J. and Dashwood, R. 1997. Inhibitory activity of green and black tea in a free radical-generating system using 2-amino-3-methylimidazo[4,5-f]quinoline as substrate. Jpn. J. Cancer Res. 88: 553–8.

Hernaez, J., Xu, M. and Dashwood, R. 1997. Effects of tea and chlorophyllin on the mutagenicity of N-hydroxy-IQ: studies of enzyme inhibition, molecular complex formation, and degradation/scavenging of active metabolites. Environ. Mol. Mutagen. 30: 468–74.

Hertog, MG., Feskens, El., Hollman, PC., Katan, MB. and Kromhout, D. 1993. Dietary antioxidant flavonoids and risk of coronary heart disease: the Zutphen Elderly Study. Lancet. 342: 1007–11.

Hour, TC., Liang, YC., Chu, IS. and Lin, JK. 1999. Inhibition of eleven mutagens by various tea extracts, (–)epigallocatechin-3-gallate, gallic acid and caffeine. Food Chem. Toxicol. 37: 569–79.

Jensson, H., Eriksson, L. and Mannervik, B. 1985. FEBS Letters 187: 115–120.

Katiyar, SK., and Mukhtar, H. 2001. Green tea polyphenol (–)-epigallocatechin-3-gallate treatment to mouse skin prevents UVB-induced infiltration of leukocytes, depletion of antigen-presenting cells, and oxidative stress. J. Leukoc. Biol. 69: 719–26.

Katiyar, SK., and Mukhtar, H. 1997. Inhibition of phorbol ester tumor promoter 12-0-tetradecanoylphorbol-13-acetate-caused inflammatory responses in SENCAR mouse skin by black tea polyphenols. Carcinogenesis 18: 1911–6.

Klaunig, J. and Yong, X. 1998. The role of oxidative stress in chemical carcinogenesis. Environmental Health Perspectives 106: 289–295.

Lea, MA., Xiao, Q., Sadhukhan, AK., Cottle, S., Wang, ZY. and Yang, CS. 1993. Inhibitory effects of extracts and (–)-epigallocatechin gallate on DNA synthesis and proliferation of hepatoma and erythroleukemia cells. Cancer Lett. 68: 231–6.

Leanderson, P., Faresjo, AO. and Tagesson C. 1997. Green tea polyphenols inhibit oxidant-induced DNA strand breakage in cultured lung cells. Free Radic. Biol. Med. 23: 235–42.

Lee, WM. 1995. Drug-induced hepatotoxicity. N. Engl. J. Med. 333: 1118–1127.

Liang, YC., Chen, YC. and Lin, YL. 1999. Suppression of extracellular signals and cell proliferation by the black tea polyphenol, theaflavin-3,3'-digallate. Carcinogenesis 20: 733–6.

Liang, YC., Lin-shiau, SY., Chen, CF. and Lin, JK. 1997. Suppression of extracellular signals and cell liferation through EGF receptor binding by (–)-epigallocatechin gallate in human A431 epidermoid carcinoma cells. J. Cell. Biochem. 67: 55–65.

Lin, JK.and Liang, YC. 2000. Cancer chemoprevention by tea polyphenols. Proc. Natl. Sci. Counc. 24: 1–13.

Lin, YL. and Lin JK. 1997. (–)-Epigallocatechin-3-gallate blocks the induction of nitric oxide synthase by down-regulating lipopolysaccharide-induced activity of transcription factor nuclear factor-*kappaB*. Mol. Pharmacol 152: 465–72.

Lin, YL., Juan, IM., Chen, YL., Liang, YC. and Lin, JK. 1996. Composition of polyphenols in fresh tea leaves and association of their oxygen-radical absorbing capacity with antiproliferative action in fibroblast cells. J. Agric. Food Chem. 44: 1387–1394.

Lotito, SB. and Fraga, CG. 1998. (+)-Catechin prevents human plasma oxidation. Free Radic. Biol. Med. 24: 435-41.

Lu, J., Ho, CT., Ghai, G. and Chen, KY. 2000. Differential effects of theaflavin monogallates on cell growth, apoptosis, and Cox-2 gene expression in cancerous versus normal cells. Cancer Res. 60: 6465–71.

Manson, M. 2003. Cancer prevention—the potential for diet to modulate molecular signaling. TRENDS in Molecular Medicine 9: 11–18.

Marx, J. 2001. Cancer research. Preventing hair loss from chemotherapy. Science 291: 25–26.

Masuda, M., Suzui, M. and Weinstein, IB. 2001. Effects of epigallocatechin-3-gallate on growth, epidermal growth factor receptor signaling pathways, gene expression, and chemosensitivity in human head and neck squamous cell carcinoma cell lines. Clin. Cancer Res. 7: 4220–9.

Miller, NJ., Castelluccio, C., Tijburg, L. and Rice-Evans C. 1996. The antioxidant properties of theaflavins and their gallate esters—radical scavengers or metal chelators? FEBS Lett. 392: 40–44.

Moncada, S. 1992. Nitric oxide gas: mediator, modulator, and pathophysiologic entity. J. Lab. Clin. Med. 120: 187–91.

Nanjo, F., Mori, M., Goto, K.and Hara, Y. 1999. Radical scavenging activity of tea catechins and their related compounds. Biosci. Biotechnol. Biochem. 63: 1621–3.

Norbury, CJ. and Hickson, ID. 2001. Cellular responses to DNA damage. Annu. Rev. Pharmacol. Toxicol. 41: 367–401.

Oltvai, ZN., Milliman, CL. and Korsmeyer, SJ. 1993. Bcl-2 heterodimerizes in vivo with a conserved homolog, Bax, that accelerates programmed cell death. Cell. 74: 609–19.

Pannala, AS., Rice-Evan, CA., Halliwell, B. and Singh, S. 1997. Inhibition of peroxynitrite-mediated tyrosine nitration by catechin polyphenols. Biochem. Biophys. Res. Commun. 232: 164–8.

Pawelec, G., Zeuthen, J. and Kiessling, R. 1997. Escape from host-antitumor immunity. Crit. Rev. Oncog. 8: 111–41.

Phillips, DH. 2001. Understanding the genotoxicity of tamoxifen? Carcinogenesis 22: 339–49.

Ray, PK. and Das, T. 1998. Molecular adaptation to toxic chemicals and drugs. In: Advances in Organ Biology (Jai Press Inc), 6: 255–69.

Ray, PK., Das, T. and Sa, G. 2001. How to activate intrinsic stress resistance mechanisms to obtain therapeutic benefit. Protein adaptation and signal transduction (edited by K.B. Storey and J.M. Storey). Elsevier Science B.V. chapter. 14: 195–202.

Rice-Evans, C. and Burdon, R. 1993. Free radical-lipid interaction and their pathological consequences. Prog. Lipid Res. 32: 71–110.

Sano, M., Takahashi, Y., Yoshino, K., Shimoi, K., Nakamura, Y., Tomita, I., Oguni, I. and Konomoto, H. 1995. Effect of tea (Camellia sinensis L.) on lipid peroxidation in rat liver and kidney: a comparison of green and black tea feeding. Biol. Pharm. Bull. 18: 1006–8.

Sell, S. 1993. Cellular origin of cancer: dedifferentiation or stem cell maturation arrest? Environ. Health Perspect. 101: 15–26.

Serafini, M., Ghiselli, A. and Ferro-Luzzi, A. 1996. In vivo antioxidant effect of green and black tea in man. Eur. J. Clin. Nutr. 50: 28–32.

Sogn, J. 1998. Tumor immunology; The glass is half full. Immunity 9: 757–63.

Sohn, OS., Surace A, Fiala ES, Richie JP Jr, Colosimo S, Zang, E. and Weisburger, JH.1994. Effects of green and black tea on hepatic xenobiotic metabolizing systems in the male F344 rat. Xenobiotica 24: 119–27.

Sosin, M. and Handa, S. 2003. Low dose methotrexate and bone marrow suppression. BMJ 326: 266–7.

Stadler, WM., Yogelzang, NJ., Amato, R., Sosman, J., Taber, D., Liebowitz, D. and Yokes, EE. 2000. Flavopiridol, a novel cyclin-dependent kinase inhibitor, in metastatic renal cancer: a University of Chicago Phase II Consortium Study. J. Clin. Oncol. 18: 371–75.

Steele, T A. 2002. Chemotherapy-induced immunosuppression and reconstitution of immune function. Leuk. Res. 26: 411–14.

Stremmel, C., Klein, P. and Hohenberger W. 2002. Immune tolerance and active suppression in oncology. Immunological principles and therapeutic options. Chirurg. 73: 255–61.

Tuting, T., Storkus, WJ. and Lotze MT. 1997. Gene-based strategies for the immunotherapy of cancer. J. Mol. Med. 75: 478–91.

Van Loveren, H, Steerenberg, PA. and Vos, JG. 1995. Early detection of immunotoxicity: from animal studies to human biomonitoring. Toxicol. Lett. 77: 73–80.

Wang, ZY., Das, M., Bickers, DR. and Mukhtar, H. 1988. Interaction of epicatechins derived from green tea with rat hepatic Cytochrome P-450. Drug Metab. Dispos. 16: 98–103.

Weisburger, JH., Hara, Y., Dolan, L., Luo, FQ., Pittman, B. and Zang, E. 1996. Tea polyphenols as inhibitors of mutagenicity of major classes of carcinogens. Mutat. Res. 371: 57–63.

Wojtowicz-Praga, S. 1997. Reversal of tumor-induced immunosuppression: a new approach to cancer therapy. J. Immunother. 20: 165–77.

Wolff, JE., Hauch, H., Kuhl J, Egeler, RM. and Jurgens, H. 1998. Dexamethasone increases hepatotoxicity of MTX in children with brain tumors. Anticancer Res. 18: 2895–9.

Wu, PF., Chiang, TA., Wang, LF., Chang, CS. and Ko, YC. 1998. Nitro-polycyclic aromiatic hydrocarbon contents of fumes from heated cooking oils and prevention of mutagenicity by catechin. Mutat. Res. 403: 29–34.

Yang, CS., Chung, JY., Yang, GY., Li, C., Meng, X. and Lee, MJ. 2000. Mechanisms of inhibition of carcinogenesis by tea. Biofactors 13: 73–9.

Yang, CS., Yang, GY., Landau, JM., Kim, S. and Liao, J. 1998. Tea and tea polyphenols inhibit cell hyperproliferation, lung tumorigenesis, and tumor progression. Exp. Lung Res. 24: 629–39.

Yen, GC. and Chen, HY. 1996. Relationship between antimutagenic activity and major components of various teas. Mutagenesis 11: 37–41.

Zhang, A., Zhu, QY., Luk, YS, Ho, KY., Fung, KP. and Chen ZY. 1997. Inhibitory effects of jasmine green tea epica Zhantechin isomers on free radical-induced lysis of red blood cells. Life Sci. 61: 383–94.

Zhang, C., Gong, Y., Ma, H., An, C. and Chen, D. 2001. Reactive oxygen species involved in trichosanthin-induced apoptosis of human choriocarcinoma cells. Biochem. J. 355: 653–61.

Zvetkova E, Wirleitner B, Tram NT, Schennach H, and Fuchs D. 2001. Aqueous extracts of *Crinum latifolium* (L.) and *Camellia sinensis* show immunomodulatory properties in human peripheral blood mononuclear cells. Int. Immunopharmacol. 12: 2143–50.

Apoptosis — A genetically determined process of cell self-destruction that is marked by the fragmentation of nuclear DNA, is activated either by the presence of a stimulus or by the removal of a stimulus or suppressing agent, is a normal physiological process eliminating DNA-damaged, superfluous, or unwanted cells (as immune cells targeted against the self in the development of self-tolerance or larval cells in amphibians undergoing metamorphosis), and when halted (as by genetic mutation) may result in uncontrolled cell growth and tumor formation, also called *programmed cell death.*

Metastasis:

1. Transmission of pathogenic microorganisms or cancerous cells from an original site to one or more sites elsewhere in the body, usually by way of the blood vessels or lymphatics.

2. A secondary cancerous growth formed by transmission of cancerous cells from a primary growth located elsewhere in the body.

Malignancy:
1. (medicine) a malignant state; progressive and resistant to treatment and tending to cause death [syn: *malignance*]
2. quality of being disposed to evil; intense ill will [syn: *malignity, malignance*] [ant: *benignity*]

Immune system:
1. The bodily system that protects the body from foreign substances, cells, and tissues by producing the immune response and that includes especially the thymus, spleen, lymph nodes, special deposits of lymphoid tissue (as in the gastrointestinal tract and bone marrow), lymphocytes including the B cells and T cells, and antibodies.
2. The integrated body system of organs, tissues, cells, and cell products such as antibodies that differentiates self from nonself and neutralizes potentially pathogenic organisms or substances.

Tumor: An abnormal growth of tissue resulting from uncontrolled, progressive multiplication of cells and serving no physiological function; a neoplasm. A swollen part; a swelling.

Immunosuppression: Lowering the body's normal immune response to invasion by foreign substances; can be deliberate (as in lowering the immune response to prevent rejection of a transplanted organ) or incidental (as a side effect of radiotherapy or chemotherapy for cancer).

Mutation: As now employed (first by de Vries), a sudden variation (the offspring differing from its parents in some well-marked character or characters) as distinguished from a gradual variations in which the new characters become fully developed only in the course of many generations. The occurrence of mutations, and the hereditary transmission, under some conditions, of the characters so appearing, are well-established facts; whether the process has played an important part in the evolution of the existing species and other groups of organisms is a disputed question. (b) The result of the above process; a suddenly produced variation.

Mutagen: An agent, such as a chemical, ultraviolet light, or a radioactive element, that can induce or increase the frequency of mutation in an organism.

Mutagenicity:	An agent, such as a chemical, ultraviolet light, or a radioactive element, that can induce or increase the frequency of mutation in an organism.
Carcinogen:	A cancer-causing substance or agent.
GPT ***abbr.*:**	glutamic-pyruvic transaminase
ALT:	Alanine aminotransferase; an enzyme in serum and body tissues that catalyzes the transfer of amino acid groups from L-alanine to 2-ketoglutarate or the reverse, thus allowing nitrogen to be excreted or incorporated into other compounds.
pH:	A measure of the acidity or alkalinity of a solution, numerically equal to 7 for neutral solutions, increasing with increasing alkalinity and decreasing with increasing acidity. The pH scale commonly in use ranges from 0 to 14. [p(otential of) h(ydrogen).]

Chapter

8

Black Tea as an Antidote for Cigarette Smoke-induced Oxidative Damage of Proteins

Shuvojit Banerjee, Anup Misra, Dhruba J. Chattopadhyay and Indu B. Chatterjee

Dr. B. C. Guha Centre for Genetic Engineering and Biotechnology and the Department of Biochemistry University College of Science, Kolkata-700 019, India

ABSTRACT

Cigarette smoke is a complex mixture of about 4000 chemicals in the form of gases and particulate fraction (tar) containing nicotine, an addictive drug, and a host of other toxic chemicals, including nitric oxide, phenols, carcinogens as well as free radicals. The hazardous effect of smoking is a global public health problem of great concern. About one-third of the world's population are direct or indirect (passive) smokers. In spite of statutory warnings and anti-smoking campaigns, smoking continues. So, a practical approach is to find a way to prevent the harmful effects of smoking. One of the prominent deleterious effects of cigarette smoke (CS) is the oxidative damage of biological macromolecules.

As tea polyphenols have strong antioxidant properties, we addressed the question whether it can prevent cigarette smoke-induced oxidative damage. In this volume, we have studied the effect of black tea (BT) on cigarette smoke-induced oxidative damage of proteins both *in vitro* and *in vivo*. The results indicate that BT prevents cigarette smoke induced oxidation of bovine serum albumin (BSA) and oxidative degradation of guinea pig lung, liver and heart microsomal proteins.

The protein oxidation and its prevention by BT were measured by immunoblotting of the dinitrophenylhydrazone derivatives of the protein carbonyls followed by densitometric scanning. Protein degradation was demonstrated by sodium dodecylsulphate polyacrylamide gel electrophoresis. The prevention of CS-induced protein oxidation by BT was accounted by its proportional contents of the principal flavonols, namely, theaflavins, thearubigins and catechins.

BT prevents microsomal protein degradation by preventing oxidative modification of the proteins. The antioxidant effect of BT was similar to that of green tea. The *in vitro* results were corroborated by the observations *in vivo*. Guinea pigs exposed to CS and given water as the drink showed extensive oxidation accompanied by loss of microsomal proteins of lung, liver and heart to the extent of 39%, 40% and 30%, respectively. When water was replaced by black tea infusion as the drink, the oxidation was prevented by approximately 90% in the lung, 97% in the liver and 70% in the heart and the protein loss was prevented by about 92% in the case of lung, 98% in the liver and 90% in the heart.

These results, if extrapolated to humans, would indicate that regular intake of tea may protect the smokers from cigarette smoke-induced oxidative damage and consequent degenerative diseases.

INTRODUCTION

Cigarette smoking is a leading cause of preventable death in the world. The combination of a highly addictive pharmacologically active substance – nicotine and an array of noxious chemicals – drastically affect the health. Cigarette smoke (CS) damages all parts of the body, gradually and insidiously, in a number of different ways. CS causes lung cancer and other malignancies, chronic bronchitis and emphysema, coronary heart disease, stroke, cerebrovascular accident, peripheral vascular disease, peptic ulcer disease, macular degeneration, cataract, damage to foetus and many other harmful effects (Bartecchi et al., 1994; Frank, 1993; Shah and Helfant, 1998; Sherman, 1991; U.S. Surgeon General's Report, 1985.). One of the prominent hazardous effects of CS is the oxidative damage of biological macromolecules including proteins and DNA (Reznick et al., 1992; Kiyosawa, 1996; Panda et al., 1999; Panda et al., 2000). The oxidative damage is produced by stable oxidants present in the aqueous extract of CS (Panda et al., 1999). The stable oxidants are very probably long-lived free radicals present in tar (Panda et al., 1999; Pryor et al., 1986; Panda et al., 2001). We reported before that the stable oxidants in CS cause oxidation of plasma proteins and extensive oxidative degradation of guinea pig lung tissue microsomal proteins (Panda et al., 1999). Because CS causes extensive oxidative damage and tea polyphenols have strong antioxidant properties, we addressed the question whether tea can prevent CS-induced oxidative damage of biological macromolecules. In fact, there is some evidence to suggest that green tea has chemopreventive effects in cigarette smokers (Shim et al., 1995; Lee et al., 1997).

Tea (*Camellia sinensis*) is one of the most widely consumed beverages in the world after water. Although about 80% of the tea produced in the world is consumed as black tea (BT) and only 20% as green tea (GT) (Krisnamoorthy, 1991) most of the researches were carried out with GT. GT contains about 30% catechins (Lee et al., 2000). Catechins are monomeric polyphenols, the principal component of which is epigallocatechin gallate (EGCG). Catechins are powerful antioxidants and effective for scavenging reactive oxygen species such as superoxide, hydrogen peroxide and hydroxyl radicals (Katiyar et al., 1996). Based on epidemiological data and experiments on animal experiments, GT has been suggested to prevent CS-induced cancers of the lung, oral cavity and oesophageus (Weisburger et al., 1999). Most of the cancer preventive effects of GT are mediated by EGCG and other catechins, through inhibition of oxidative reactions (Lin et al., 2000). It has been shown that GT polyphenols inhibit both CS and H_2O_2-induced DNA breakdown (Leanderson et al., 1997). In contrast to GT, BT has been assumed to have much less antioxidant property because of its lower content of monomeric polyphenols (Katiyar et al., 1996; Wang et al., 1993). As a result, not much research on tea has been carried out with BT. The leaves of the plant *C. sinensis* contain not only about one-third of the dry weight as polyphenols, but also the enzyme, polyphenol oxidase. When the leaves are harvested, withered and immediately steamed or heated, the polyphenol oxidase is inactivated, yielding green tea. However, if the leaves at harvest are withered, rolled and crushed, the polyphenol oxidase liberated converts the polyphenols into black tea within a 60–90-minute reaction time, after which the product is dried in a stream of hot air. Therefore, black teas stem from the biochemical oxidation of the tea polyphenols. In the trade, this process is called "fermentation". The polyphenols in green tea are catechin derivatives, and the main product is the polyphenol epigallocatechin gallate. Biochemical oxidation of these catechins yields the polyphenols in black tea, such as theaflavins (TF) and thearubigins (TR), a fairly complex number of polymeric chemicals, that retain potent physiological and biochemical properties similar to the attributes of epigallocatechin gallate in green tea leaves (Lin et al., 2000). TFs are a mixture of TF and TF-gallates. However, TRs are more complex oligomeric flavonols; those vary greatly in molecular weight ranging from 1,000 to 40,000 Da (Lee et al., 2000). The contents of TF and TR vary with the species of tea and the process of "fermentation". Nevertheless, TRs are the most abundant phenolic fractions in BT. BT contains about 1-4% TF, 10-20% TR and also catechins (CT) (Lee et al., 2000; Yang et al., 1993). Recently, initiated has been initiated with BT and its characteristic constituents, TF and TR. However, there are conflicting reports on the comparative beneficial effects of GT and BT with particular reference to the chemoprevention of cancer (Jankun et al., 1997; Yang, 1997). Other reports indicate that TF in BT and CT in GT are equally effective antioxidants (Lin and Liang, 2000; Leung et al., 2001; Wiseman et al., 1997; Weisburger, 1999; Grinberg et al., 1997; Zhang et al., 1997; Haldar and Bhaduri, 1998). One observation reported that the extracts from BT scavenged hydrogen peroxide more potently than those from GT (Wei et al., 1999). In this chapter we demonstrate that BT and its constituent polyphenols, namely TF and TR, prevent CS-induced oxidative damage of proteins both *in vitro* and *in vivo*.

We and others have reported that oxidative modification of proteins by stable oxidants of CS introduces carbonyl groups into side chain amino acids (Reznick et al., 1992; Panda et al., 1999; Panda et al., 2000; Panda et al., 2001). The formation of the carbonyl groups can be measured spectrophotometrically after converting these into DNP-derivative by reaction with 2,4-dinitrophenylhydrazine (Reznick et al., 1992; Levin, 1990). However, attempts to quantify the preventive effects of BT on CS-induced oxidation of proteins by spectrophotometric assay of the DNP-derivative produce erroneous results. This is because BT interacts with proteins and the product gives absorbance at 360-390, the range of the wavelength at which the DNP derivative is measured. In this volume we have used the techniques of immunobloting accompanied by densitometric scanning for quantifying the preventive effect of black tea and its components on CS-induced oxidative damage of proteins.

MATERIALS AND METHODS

Reagents Unless otherwise mentioned, in all experiments CTC black tea was used. The source of both black and green tea was West Bengal Tea Development Corporation, Kolkata, India. CT, epigallocatechin gallate (EGCG), BSA and tea extracts were purchased from Sigma Chemical Co., St. Louis, Missouri, USA. Tea extract containing about 80% TF was used as a source of TF. TR was prepared by the method described below. All other chemicals were of analytical grade.

Preparation of Tea Infusion One gram black or green tea was added to 10 ml of boiling water, brewed for 5 min, cooled to room temperature and filtered. The filtrate has been designated as BT or GT infusion.

Measurement of Theaflavins and Catechins in BT Diluted samples of BT infusion were analysed by HPLC in a Shimadzu instrument consisting of a SPD-10A VP UV-Vis detector, LC-10A VP dual pump systems, attached to a chromatopac C-R6A. Individual polyphenols were used as standards. A Lichrospher$^{(R)}$ 100 RP-18 encapped (5 μm; 250 mm × 4 mm) column (Merck) was used. CT and TF were detected at 270 nm using a binary gradient cycle as described in details by Lee et al. (1997).

Preparation of Polyphenol-free BT Infusion BT infusion as prepared above was treated with polyvinyl-polypyrolidone (30 mg per ml of infusion) batch wise two times until the filtrate was faint yellow. The filtrate was then passed through a bed of activated charcoal to remove the residual polyphenols as well as other small molecular weight components. The effluent from the charcoal bed was practically colourless without any absorbance at 270 nm.

Preparation of TR TRs were isolated from BT following a combination of methods of Subramanian et al. (1999) and Hara (1995). BT (6 gram) was boiled in 50 ml of 0.01M sodium acetate (pH 5.0) for 10 min, cooled and filtered. The filtrate was extracted successively with equal volumes of chloroform, methyl isobutyl ketone and ethyl acetate. The organic layers were discarded and the aqueous layer was extracted with butanol followed by lyophilisation. The residual dark orange powder was used as TR, the content of which was measured spectrophotometrically at 270 nm (Serafini et al., 1996), as described below.

Measurement of TR in BT Spectroscopic analysis showed that BT infusion and all the individual polyphenols, namely, TF, TR and CT have λ_{max} at 270 nm. We calculated the TR content of BT by deducting combined A_{270} due to TF and CT from A_{270} of BT infusion. As stated above, TF and CT in BT were estimated separately by HPLC analysis.

Preparation of Cigarette Smoke Solution (CS-solution) Smoke from an Indian commercial filter-tipped cigarette (74 mm) with a tar content of 22 mg was dissolved in one ml phosphate buffer (50 mmole/l), pH 7.4 as described before (Panda et al., 1999). The pH of the yellow coloured solution was adjusted to 7.4 by the addition of sodium hydroxide solution (80 gram/l) followed by filtration through 0.22 μm Millipore filter.

Exposure of Guinea Pigs to Cigarette Smoke (CS) Male short hair guinea pigs weighing 400-500 gm obtained from a local animal supplier, were used. All animal care procedures met the National Institute of Health guidelines (National Research Council (1985) Guide for care and use of laboratory animals. Publication No. 85-23 (rev.) National Institutes of Health. The guinea pigs were kept on ascorbate free diet to minimise the ascorbate level of plasma and tissues, because ascorbate is a potential inhibitor of CS-induced oxidative damage of proteins, which would otherwise counteract the damaging effect of CS (Panda et al., 1999; Panda et al., 2000). The diet contained in gm/kg: wheat flower 700; casein 200; sucrose 80; USP XVII salt mixture, 10; AOAC vitamin mixture, 10 (Nandi et al., 1973). After feeding ascorbate-free diet for 7 days, the guinea pigs were subjected to CS exposure from 5 cigarettes/animal/day for further 7 days in a smoke chamber as described above (Panda et al., 2000). The duration of exposure was 1 minute for each puff. Altogether 2 puffs per cigarette were given, allowing the guinea pigs one minute rest in a smoke-free atmosphere to breathe air between each puff. The gap between exposure to one cigarette and the next was one hour. This was done as an attempt to simulate the mode of smoking by humans. Pair-fed sham controls were subjected to air exposure instead of CS exposure. The guinea pigs were divided into the following experimental groups. Each group contained 4 animals.

Gr.I. Sham controls. The guinea pigs were exposed to air instead of CS and given water as the drink.

Gr.II. The guinea pigs were exposed to CS and given water as the drink.

Gr.III. The animals were exposed to CS and given black tea infusion (20 gram/l) as the drink instead of water. The tea infusion was freshly prepared and replaced every morning and evening.

Gr.IV. The guinea pigs were exposed to air and given black tea infusion as the drink.

All the guinea pigs were pair-fed individually with respect to animals of Gr. II. The amount of diet ingested and fluid intake by the Gr. II guinea pigs on the previous day were given to the other groups. After feeding ascorbate-free diet for 7 days following exposure to CS/air for further 7 days, both the sham controls and the CS-exposed guinea pigs, after overnight food-deprivation, were killed on the following day by diethyl ether and lung, liver and heart tissue excised immediately and processed without delay. Oxidative damage of the tissue microsomal proteins, without any further treatment, was assayed by immunoblotting and sodium dodecylsulphate polyacrylamide gel electrophoresis (SDS-PAGE). SDS-PAGE was carried out as before (Panda et al., 1999). The method of immunoblotting is given below. Protein was estimated by the method of Lowry et al. (1951) as described before (Panda et al., 1999).

Preparation of Microsomes Portions of the lung, liver and heart from the guinea pigs of each group were pooled, homogenized and microsomes prepared and washed as described earlier (Panda et al., 1999). The washed microsomes were dispersed in potassium phosphate buffer (50 m mole/L), pH 7.4, so that 1 mL of the dispersion was equivalent to 1 gram tissue.

Incubation System Incubation mixture contained BSA (1 mg) or tissue microsomal suspension (equivalent to 1 mg of protein) with or without 50 μl of CS-solution (equivalent to one-twentieth of a cigarette) in the presence or absence of 50 μl of tea infusion (equivalent to 5 mg tea) or its components in a final volume of 200 μl of potassium phosphate buffer (50 m mole/l), PH 7.4. Incubation was carried out at 37°C with shaking for 1 hour or as described. In studies with whole phase cigarette smoke, the smoke was directly released (without bubbling) on 1 ml of potassium phosphate buffer (50 m mole/l), pH 7.4 containing 20 mg of BSA with gentle shaking. Fifty μl of this solution containing 1 mg of the smoke-treated protein was added to 150 μl potassium phosphate buffer (50 m mole/l), pH 7.4, to make the final incubation volume to 200 μl.

Production of Dinitrophenylhydrazone Derivative of Oxidized Protein (DNP-Protein). After incubation as stated above for 1 h , 5 μl of the incubation mixture was mixed with 5 μl of SDS (120 gram/l) solution. After 5 minutes, 10 μl of 2,4-dinitrophenyl-hydrazine (DNPH) solution (10 m mole/l) in HCl (2 moles/l) were added and the reaction was allowed to continue for 25 min followed by addition of 5 μl of neutralizing solution as per Oxyblot™ protein oxidation detection kit mentioned below.

Measurement of Protein Carbonyl by Immunoblotting of the DNP-Protein Twenty-five μl of the DNP derivative of BSA or microsomal proteins containing 25 μg of protein as described above, were separated by SDS-PAGE (100 gram/l) followed by Western blotting using PVDF membrane according to the protocol of Oxyblot™ protein oxidation detection kit (Catalogue #S7150 of Intergen company, NY, USA). The membrane was then incubated with primary antibody, specific to the DNP-moiety. This step was followed by incubation with a secondary antibody conjugated with horseradish peroxidase (goat anti-rabbit IgG) directed against the primary antibody. The membrane was then treated with chemiluminescent reagents (luminol and enhancer) and the chemiluminescence was detected by autoradiography. Intensity of the blots produced by autoradiography was quantified by densitometric scanning.

Measurement of Ascorbic Acid in BT Ascorbic acid was measured by HPLC analysis of samples of tea infusion in a Shimadzu instrument as described above under measurement of TF and CT in BT. The column used was Lichro CART 250-4 NH_2 (Merck); the mobile solvent, acetonitrile: potassium dihydrogen phosphate solution containing 50 m moles/l (75:25,v/v); flow rate, 1.5 ml/minute. Ascorbic acid was detected at 254 nm; t_r was 6.1 minutes; limit of detection 500 pg.

Statistics Results were analysed by two-way ANOVA. Classification factors were CS-treatment and effect of BT infusion, as well as their interaction. Bonferroni post hoc test was used when the F-values were significant. Means were considered significantly different at $P < 0.05$. Data are expressed as means ± SD.

RESULTS

Prevention of CS-induced BSA Oxidation by Black Tea (BT) and its Constituent Polyphenols Figure 8.1 (lane 2) shows that after incubation of 1 mg of BSA with 50 µl of CS-solution followed by reaction with DNPH and immunoblotting, extensive formation of the DNP-derivative occurs, which is a measure of the extent of carbonyl groups produced by oxidation of the side chain amino acids of BSA. Spectroscopic assay of the DNP-derivative (Panda et al., 1999) shows that under similar conditions the amount of protein carbonyl formed is 10±0.5 n moles. When BSA was incubated with CS-solution in the presence of 50 µl of BT, the oxidation of protein was inhibited about 70 % (Fig. 8.1, lane 3), as evidenced from the comparative scanning of the blots. Incubation of BSA with BT alone did not produce any oxidation of protein. The inhibition by BT infusion was not affected by

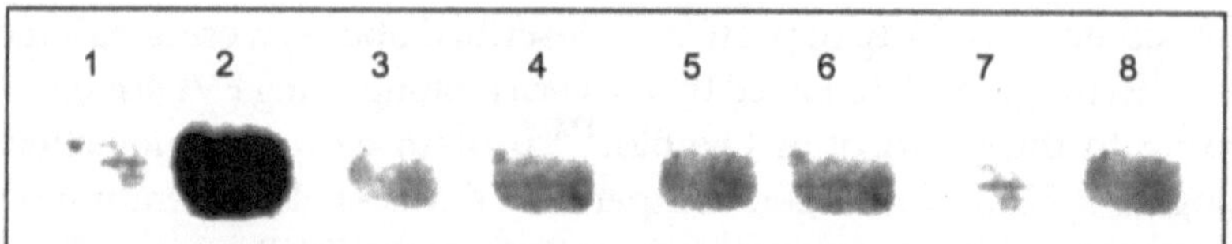

Fig. 8.1 Immunoblot of the DNP-derivatives of BSA incubated with or without CS in the presence and absence of BT and its components. The incubation mixture contained BSA, 1 mg; CS-solution 50 µl; BT or GT, 50 µl in a final volume of 200 µl of potassium phosphate buffer (50 m mole/l), pH 7.4 ; incubated for 1 h . After incubation, 5 µl of the incubated mixture was subjected to the production of DNP-derivative followed by immunoblotting as described under Materials and Methods. Lane 1, BSA; lane 2, BSA+CS; lane 3, BSA + CS +BT; lane 4, BSA+CS+TF; lane 5, BSA+ CS+ TR ; lane 6, BSA + CS + CT ; lane 7, BSA + CS + TF TR + CT ; Lane 8, BSA +CS + GT. The amounts of components used were TF, 50 µg ; TR, 890 µg; CT, 300 µg. Similar percentage of inhibition was observed from four independent experiments; SD < 10%.

addition of desferrioxamine (DFO, 20 µmole/l) to the incubation mixture, indicating that the inhibition was independent of free iron. Protection of CS-induced BSA oxidation was also observed by TF, TR and CT, the principal components of BT. HPLC analysis indicated that 50 µl of the BT infusion contained 50 µg of TF which amounts to about 1% TF in the sample of BT used. When instead of 50 µl BT, 50 µg of TF were added to the incubation mixture containing 1 mg of BSA and 50 µl of CS solution, the inhibition of BSA oxidation was 50% (Fig. 8.1, lane 4). It is known that amongst the oxidized polyophenols, TRs are most abundant in BT. The amount of TR in the BT used was found to be 890 µg per 50 µl of BT infusion (approximately 17.8%; see Materials and Methods). Fig 8.1, (lane 5) shows that 890 µg of TR inhibited CS-induced BSA oxidation about 72%. HPLC analysis revealed that besides TF, 50 µL of the BT infusion used contained 300 µg CT, (approximately 6%). When used separately, 300 µg CT produced 85% inhibition (Fig. 8.1, lane 6). The results show that among the polyphenols, CT has maximum antioxidant effect when used in proportions of their occurrences in BT. However, when used in identical amounts, TF appears to be the most potential antioxidant for inhibiting CS-induced BSA oxidation. While 50 µg TF inhibited 50%, 50 µg CT inhibited only 20% and 50 µg TR was practically ineffective. When instead of the BT infusion, a mixture of 50 µg TF, 890 µg TR and 300 µg CT were used, the inhibition of BSA oxidation was almost 100% (Fig. 8.1, lane 7), which is more than that obtained with BT infusion. The data indicate that the additive effect of the constituent polyphenols is more than that when present in the BT infusion. The preventive effects of the polyphenols were confirmed by adding the individual polyphenols to polyphenol-free BT infusion. Polyphenol free-BT infusion alone failed to protect CS-induced BSA oxidation. When TF, TR, and CT, in relative proportions of their occurrences in BT as stated above, were added separately and conjointly to the polyphenol-free BT infusion, the polyphenols inhibited CS-induced BSA oxidation to a pattern similar to that obtained with individual polyphenols in the absence of BT infusion. Addition of 50 µg TF, 890 µg TR and 300 µg CT to 50 µl polyphenol-free BT infusion produced inhibition to the extent of 47%, 70% and 80% respectively. When used conjointly, the inhibition was almost 100% .When instead of BT, 50 µl of GT infusion was used, the inhibition of BSA oxidation was 72% (Fig. 8.1, lane 8). These results demonstrate that although BT contains mostly oxidized and complex condensation products of CT, both BT and GT have similar antioxidant effects against CS-induced BSA oxidation. HPLC analysis showed that the amount of EGCG present in 50 µl GT was 870 µg. When 870 µg EGCG were added to the incubation mixture, the inhibition of

CS-induced BSA oxidation was 98%, which is more than that obtained with GT infusion. This again indicates that the antioxidant effect of individual polyphenol is more than that when present in the tea infusion. Similar preventive results of BT and GT were obtained when BSA was oxidized by whole phase cigarette smoke instead of CS-solution (data not shown).

The antioxidant effect of BT was similar whether the tea was harvested in summer or autumn. In separate experiments it was observed that tea infusion prepared from orthodox Darjeeling or Assam tea had same antioxidant effects to that of CTC tea. The protection of BSA oxidation by BT was a function of the concentration of BT infusion. While 25 µl of tea gave 43% protection, 50 µl of tea inhibited about 70%, 75 µl gave 90% and 100 µl gave full protection (Fig. 8.2). With 50 µl of tea, the percent of inhibition of BSA oxidation with correspondingly increased the duration time of incubation and attained a maximum value after 1 h (Fig. 8.3).

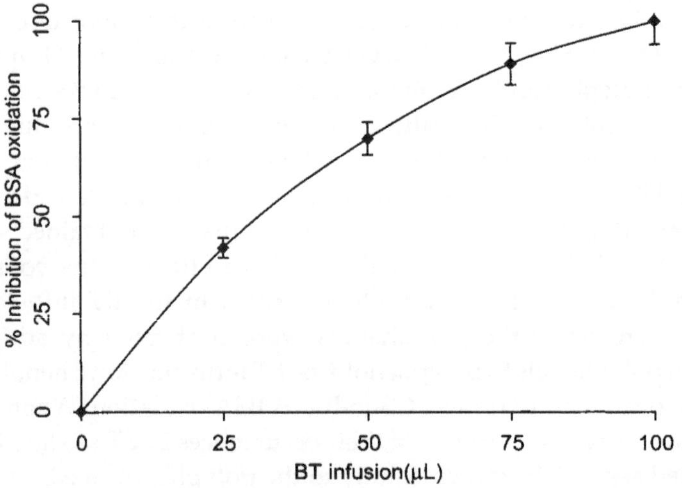

Fig. 8.2 Protection of CS-induced BSA oxidation as a function of concentration of BT infusion. Conditions are same as in Fig. 8.1. Data represent means of four independent observations; bar on each point are means ± SD: $P > 0.05$

In India, the number of people who smoke 'bidi' (made from tobacco leaf) is about four times that of cigarette smokers. Bidi smoking is also gaining popularity abroad. We have observed that 'bidi' smoke solution, like CS-solution, produces extensive oxidation of BSA, which is markedly (70%) prevented by black tea infusion, as evidenced by immunoblotting (data not shown).

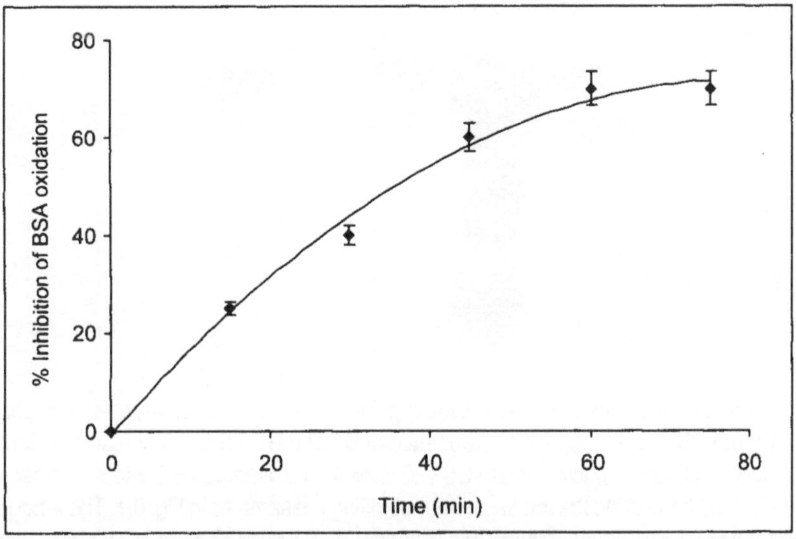

Fig. 8.3 Inhibition of CS-induced BSA oxidation as a function of time. Conditions are same as in Fig. 8.1. The amount of BT was used 50 μl. Values are means of four independent observations; bar on each point are means ± SD; $P < 0.05$.

It has been reported previously that ascorbic acid protects cigarette smoke-induced BSA oxidation (Panda et al., 1999). However, the inhibition of CS-induced BSA oxidation by tea infusion was not due to the presence of ascorbic acid. HPLC analysis of both BT and GT infusion revealed that ascorbic acid was practically absent in the tea infusion.

Prevention of CS-induced Oxidation of Guinea Pig Lung Microsomal Proteins by BT Figure 8.4 (lane 2) shows that incubation of guinea pig lung microsomes with CS-solution, followed by reaction with DNPH and immunobloting, results in marked oxidation of the microsomal proteins. It was reported before that CS-induced microsomal protein oxidation is accompanied by increased proteolysis. So, phenylmethylsulphonyl fluoride (PMSF, 20 μg) and EDTA (2 mmole/l) were added to the incubation mixture to prevent proteolytic degradation of the oxidized proteins. Densitometric scanning shows that the protein oxidation is about 97% inhibited by BT infusion (Fig. 8.4, lane 3). Incubation of the microsomes with only black tea infusion does not produce any oxidation (Fig. 8.4, lane 4). GT produced 93% protection (Fig. 8.4, lane 5).This indicates that the antioxidant effect of BT and GT are similar.

Cigarette Smoke-induced Degradation of Microsomal Proteins and Protection by BT. Figure 8.5 (lane 1) shows that compared to untreated microsomes,

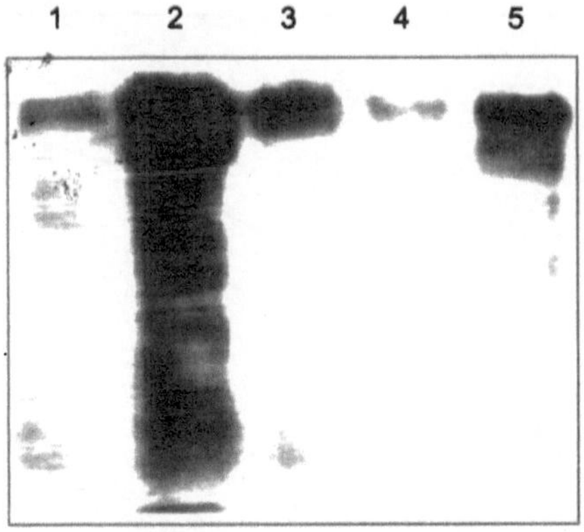

Fig. 8.4 Immunoblot of the DNP-derivatives of guinea pig lung microsomal proteins incubated with or without CS in the presence and absence of BT and GT. Lane 1, microsomes; lane 2, microsomes + CS; lane 3, microsomes +CS +BT; lane 4, microsomes +BT; lane 5, microsomes + CS +GT. Conditions of incubation and immunoblotting are same as in Fig. 8.1. The amount of microsomes used was equivalent to 1 mg protein. CS, BT and GT were 50 µl each. Similar observations were made from four independent experiments.

incubation of guinea pig lung microsomes with CS-solution (lane 2) results in degradation of the microsomal proteins to the extent of 90%, as evidenced from densitometric scanning of the protein bands. In this case, the protease inhibitors PMSF and EDTA were absent from the incubation medium. The CS-induced protein degradation is inhibited by BT infusion to the extent of 80% (Fig. 8.5, lane 3). It has been reported earlier that CS-induced microsomal protein degradation is a two-step process: (i) oxidation of the proteins by oxidants present in CS solution and (ii) rapid proteolytic degradation of the oxidized proteins by proteases present in the microsomes (Panda et al., 1999). It has been further reported that if the oxidation of proteins is prevented, proteolytic degradation does not occur (Panda et al., 1999). Since BT inhibits CS-induced oxidation of microsomal proteins (Fig. 8.4), subsequent proteolytic degradation is prevented. Similar observations were made with guinea pig heart and liver microsomes (data not shown).

BT prevents CS-induced Oxidation of Tissue Microsomal Proteins of Guinea Pigs in vivo Figure 8.6 exhibits immunoblots of lung, liver and heart microsomal proteins of guinea pigs. In each panel, lane 1 represents air-exposed animals; lane 2, after exposure to CS and given water as the drink; lane 3, after exposure to CS and given BT infusion as the drink. When guinea pigs were exposed to CS for 7 days at an exposure rate of 5 cigarettes per guinea pig per day and given water as the drink, extensive oxidation of the microsomal proteins of lung, liver and heart occurred (Fig. 8.6, lane 2). There was

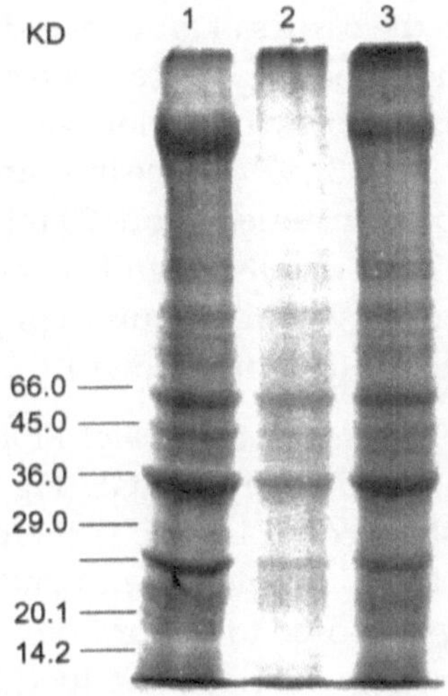

Fig. 8.5 SDS-PAGE of guinea pig lung microsomal proteins incubated with or without CS in the presence and absence of BT. Microsomal suspension (1 mg protein) was incubated at 37°C for 2 h in a final volume of 200 μl of potassium phosphate buffer (50 m mole/l). After incubation, 40 μl of the incubation mixture were subjected to SDS-PAGE using 7.5% gel. The gels were stained with Comassie brilliant Blue-250. Lane 1, microsomes incubated without CS; lane 2, microsomes incubated with 50 μl CS solution in the absence of BT; lane 3, microsomes incubated with 50 μl CS-solution in the presence of 50 μl BT. The amounts of components used are as in Fig. 8.4. Similar observations were made from four independent experiments.

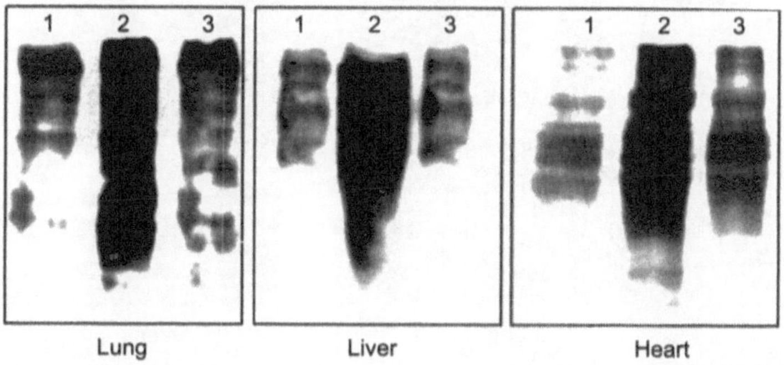

Fig. 8.6 Immunoblots of the DNP-derivatives of lung, liver and heart microsomal proteins of guinea pigs exposed to air or CS with or without BT infusion given as the drink. Twenty five μg protein equivalent of microsomes, isolated from CS-exposed guinea pigs, were subjected to without any further treatment to the production of the DNP-derivative followed by immunoblotting as mentioned in Fig. 8.1. In each panel, lane 1 exhibits microsomes from guinea pigs exposed to air (sham control) and given water as the drink; lane 2, microsomes from guinea pigs exposed to CS and given water as the drink; lane 3, microsomes from guinea pigs exposed to CS and given BT infusion as the drink. Similar observations were made from four independent experiments.

practically no oxidation in the tissues of air-exposed animals (Fig. 8.6, lane 1; sham controls). When the guinea pigs were given BT infusion as the drink instead of water, CS-induced protein oxidation was prevented to the extent of 90% in the lung (Fig. 8.6, lane 3), 97% in the liver and 70% in the heart. On an average, each guinea pig consumed about 25 ml of BT infusion, which was equivalent to 0.5 g BT per day. No significant oxidation of the tissue microsomal proteins was observed in the guinea pigs given BT infusion as the drink without CS-exposure (data not shown).

BT Prevents Degradation of Tissue Microsomal Proteins of Guinea pigs in vivo. Protein profiles obtained by SDS-PAGE (Fig. 8.7, lane 2) show that, compared to air exposed sham controls (lane 1), there is approximately 39% loss of protein bands in the lung microsomes and 40% loss in the liver microsomes of guinea pigs exposed to CS for 7 days at an exposure rate of five cigarettes per guinea pig per day. The food intake of the animals (approximately 65±5 g/animal/day) did not decrease significantly after CS exposure. Hence, the microsomal protein loss was not due to inanition, but apparently due to oxidative degradation of proteins as a result of CS

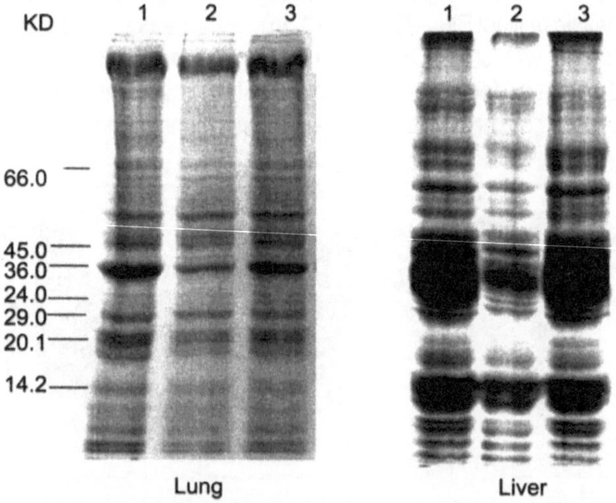

Fig. 8.7 SDS- PAGE of lung and liver microsomal proteins of guinea pigs exposed to air or CS with or without BT infusion given as the drink. Microsomes equivalent to 18 mg of tissue isolated from guinea pigs were, without any further treatment, subjected to SDS-PAGE as describe in Fig. 8.5. Electrophoresis was done using 10% gel. The proteins were stained with Comassie Brilliant Blue-250. Lane 1, guinea pigs exposed to air (sham control) and given water as the drink; lane 2, guinea pigs exposed to CS and given water as the drink; lane 3, guinea pigs exposed to CS and given BT infusion as the drink. Similar observations were made from four independent experiments.

exposure. When the CS-exposed animals were given BT infusion as the drink instead of water, the microsomal protein loss was prevented to the extent of 92% in the case of lung (Fig. 8.7, lane 3) and 98% in the case of liver, as evidenced by comparative densitometric scanning. There was also discernible loss of protein bands (≈ 30%) of the heart microsomes of the guinea pigs after exposure to CS, which was again prevented by about 90% by drinking black tea (Fig. not shown).

DISCUSSION

Oxidative damage has been projected as a prominent mechanism of the deleterious effects of CS (Reznick et al., 1992; Kiyosawa et al., 1990; Panda et al., 1999, Panda et al., 2000, Panda et al., 2001, Shalini et al., 1994, Hulea et al., 1995; Veyssier et al., 1997; Kleages et al., 1998). There have been attempts to find suitable antioxidants that can prevent CS-induced oxidative damage and the consequent degenerative diseases. We have reported before that CS-induced oxidative damage of proteins is markedly inhibited by ascorbic acid both *in vitro* and *in vivo* (Panda et al., 1999; Panda et al., 2000; Panda et al., 2001). In this volume we demonstrate that CS-induced protein oxidation is almost completely prevented by BT infusion. The antioxidant effect of BT has been found to be similar to that of GT. This indicates that although BT contains oxidized and complex condensation products of CT, the antioxidant properties are not reduced. HPLC and spectrophotometric analyses show that the sample of BT (CTC) used contain about 1% TF, 17.8% TR and 6% CT. The antioxidant property of BT is a combined effect of the flavonols. When used in relative proportions of their occurrences in BT, CT and TR appear to have more antioxidant effects than that of TF. But when used in identical amounts, TF and its gallate seem to be better antioxidant than that of TR and CT. HPLC analysis reveals that the percent distribution of TF and its gallates in the sample of BT used is TF, 27%; TF-3-monogallate, 18%; TF-3′-monogallate, 23% and TF 3,3′- digallate, 32%. The mechanism of preventive action of the BT flavonols on CS-induced oxidation of protein is not clear at present. It is known that polyphenols are reducing agents that function as antioxidants by virtue of their hydrogen donating properties of the polyphenolic hydroxyl groups as well as their transition metal chelating abilities (Jankun et al., 1997). However, addition of DFO to the incubation system did not affect the preventive effect of BT on CS-induced BSA oxidation. This shows that under the conditions TF, TR and CT of BT do not act as chelating agents and probably reduce the CS-oxidants. The CS-oxidants are probably long-lived free radicals present in aqueous solution of CS (Pryor et al., 1986). It is possible that the antioxidant flavonols in BT quench the free radicals and inactivate these. Recently, several studies have found that BT and GT offered protection against oxidative damage of red blood cells induced by a variety of inducers, e.g. H_2O_2, primaquine, phenyl hydrazine, Cu^{2+}-ascorbic acid and xanthine/ xanthine oxidase system (Grinberg et al., 1997; Zhang et al., 1997; Halder and Bhaduri, 1998). Oral administration of GT or BT inhibited lipid peroxidation of liver induced by tert-butylhydroperoxide in rats (Sano et al., 1995; Serafini et al., 1996) demonstrated that ingestion of GT and BT produces a significant increase in human plasma antioxidant capacity. Catechin polyphenols also decreased peroxynitrite-induced nitration of tyrosine and protected apolipoprotein of LDL from modification of critical amino acids (Pannala et al., 1997).

The *in vitro* observations have been corroborated by the results obtained in vivo. For *in vivo* experiments, marginal ascorbate-deficient guinea pigs were used. This is because ascorbate is a powerful antioxidant against CS-induced protein damage (Panda et al., 1999; Panda et al., 2000). When guinea pigs were exposed to CS, there was extensive oxidation of protein accompanied by protein loss of guinea pig tissue microsomes. The oxidation and protein loss was completely prevented when the animals were given BT infusion as the drink instead of water.

The degenerative diseases that have been linked to oxidative damage caused by CS are emphysema (Schraufstatter et al., 1984; Janoff, 1986; Pryor et al., 1986) atherosclerosis (Veyssier, 1997; Ueyama et al., 1998; Yokode et al., 1996; Mahhfouz et al., 1995), lung cancer and several other incidences of cancer (Asami et al., 1997; Kasai, 1989; Cross et al., 1987). Epidemiological reports and other experimental studies suggest that tea has chemo-preventive effects against cigarette smoking and tobacco use (Shim et al., 1995; Lee et al., 1997; Weisburger, 1999; Weisburger, 1997). Virtually one-third of the world's populations are direct or indirect (passive) smokers (Pianezza et al., 1998) and the hazardous effect of smoking is a global public health problem worldwide causing grave concern. Discouraging smoking would definitely be the best measure to eradicate this deleterious practice. However, in spite of the statutory warnings and antismoking campaigns, smoking continues. A practicable approach is to find the means for preventing the deleterious effects of CS. AT baseline the results presented in this paper, if extrapolated to humans, would indicate that regular intake of tea, the most widely used inexpensive beverage in the world, may protect the smokers from CS-induced oxidative damage and the consequent degenerative diseases.

ACKNOWLEDGEMENT

This research was partly supported by National Tea Research Foundation, Tea Board. We thank Dr. K. Panda for his interest in this project and Dr. A.K. Dasgupta for assisting in the densitometric scanning. We also thank Prof. S. Ganguly and HILLA tea estate, West Bengal Tea Development Corporation, for the supply of BOP (CTC) tea.

REFERENCES

Asami, S., Manabe, H., Miyake, J. and Itoh, H. 1997. Cigarette smoking induces an increase in oxidative DNA damage, 8-hydroxyguanosine in central site of the human lung. Carcinogenesis. 189: 1763-1766

Bartecchi, C.E., Mackenzie, T.D. and Schier, R.W. 1994. The human costs of tobacco use. N. Engl. J. Med. 330: 907-912

Cross, C.E. 1987. Moderator. Oxygen radicals and human disease. Ann. Intern. Med. 107: 526-545

Frank, E. 1993. Benefits of stopping smoking. West J. Med. 159: 83-87.

Grinberg, L.N, Newmark, H., Kitrossky, N., Rahamim, E., Chevion, M, and Rachmilewitz, E.A. 1997. Protective effects of tea polyphenols against oxidative damage of human red blood cells. Biochem. Pharmacol. 54: 973-978.

Haldar, J. and Bhaduri, A. N. 1998. Protective role of black tea against oxidative damage of human red blood cell. Biochem. Biophys. Res. Commun. 244: 903-907

Hara, Y. 1995. Action of tea on cardiovascular disease. Proc. International Tea-quality—Human Health Symposium. pp 16-31, Shanghai, China.

Hulea, S.A., Olinesku, K., Crocnan, D.C. and Kummerow, F.A. 1995. Cigarette smoking causes biochemical changes in blood that are suggestive of oxidative stress: a case control study. J. Environ Pathol. Toxicol. Oncol. 143-144: 173-180.

Jankun, J., Selman, S.H. and Swiercz, R. 1997. Why drinking green tea could prevent cancer. Nature 387: 561.

Janoff, A. 1986. Investigations into the biochemical mechanisms of pulmonary emphysema: effects of cigarette smoke on enzymes and anti-enzymes in the lung. Respiration 50 (suppl.): 13-25.

Kasai, H. 1989. DNA damage by oxygen radicals and carcinogenesis. Gan to Kahaku Ryoho (Japan). 163 pt 2: 459-465.

Katiyar, S.K. and Mukhtar, H. 1996. Tea and chemo-prevention of cancer. Int. J. Oncol. 8: 221-238.

Kiyosawa, H., Suko, M., Okudaira, H., Murata, K. and Nishimura, S. 1990. Cigarette smoking induces formation of 8 hydroxyguanosine, one of oxidative DNA damages in human peripheral leukocytes. Free Radic. Res. Commun. 11: 27-29.

Kleages, L.M., Murray, D.M., Brown, J.E., Cliver. S.P. and Goldenberg, R.L. 1998. Relation of cigarette smoking and dietary antioxidants with placental calcification. Am. J. Epidemiol. 1472: 127-135.

Krisnamoorthy, K.K. 1991. The nutritional and therapeutic value of tea. In proceedings of the international symposium on Tea Science, Shizuoka, Japan. pp. 1-6.

Leanderson, P., Faresjo, A.O. and Tagesson, C. 1997. Green tea polyphenols inhibit oxidant-induced DNA strand breakage in cultured lung cells. Free Radic. Biol. Med. 23: 235-242

Lee, I.P., Kim, Y. Hi. and Kang, M.H. 1997. Chemo-preventive effect of green tea against cigarette smoke-induced mutations in humans. J. Cell. Biochem. 27: S68-S75.

Lee, M-J., Prabhu, S., Meng, X., Li, C. and Yang, C. S. 2000. An improved method for the determination of green and black tea polyphenols in biomatrices by high-performance liquid chromatography with colorimetric array detection. Anal. Biochem. 279: 164-169.

Leob, L.A., Ernster, V.L., Warner, K.E. and Abbots, J. 1984. Cigarette smoke and lung cancer. Cancer Res. 44: 5940-5958.

Leung, L.K., Su,Y., Chen, R., Zhang, Z., Huang, Y. and Chen, Z.Y. 2001. Theaflavins in black tea and catechins in green tea are equally effective antioxidants. J. Nutr. 131: 2248-2251

Levin, R.L., Garl and D., Oliver, C.N., Amici, A., Climent, L., Lenz, A., Ahn, B., Shaltiel, S. and Stadtman, E.R. 1990. Determination of carbonyl content in oxidatively modified proteins. Methods Enzymol. 186: 464-478.

Lin, J-K. and Liang, Y-C. 2000. Cancer chemoprevention by tea polyphenols. Proc. Natl. Sci. Counc. Roc B. 24: 1-13.

Lowry, O.H., Rosenbrough, N.J., Farr, A.L. and Randall, R.J. 1951. Protein measurement with the folin phenol reagent. J. Biol. Chem. 193: 265-275.

Mahhfouz, M.M., Hulea, S.A. and Kummerow, F.A. 1995. Cigarette smoke increases cholesterol oxidation and lipid peroxidation of low-density lipoproteins and decreases its binding to hepatic receptors in vitro. J. Environ. Pathol. Toxicol. Oncol. 14(3, 4): 181-192.

Nandi, B.K., Majumder, A.K., Subramanian, N. and Chatterjee, I.B. 1973. Effects of large doses of vitamin C in guinea pigs and rats. J. Nutr. 103: 1688-1695

National Research Council. 1985. Guide for care and use of laboratory animals. Publication No. 85-23 (rev.) National Institutes of Health.

Panda, K., Chattopadhyay, R., Ghosh, M.K., Chattopadhyay, D.J. and Chatterjee, I. B. 1999. Vitamin C prevents cigarette smoke-induced oxidative damage of proteins and increased proteolysis. Free Radic. Biol. Med. 27: 1064-1079.

Panda, K., Chattopadhyay, R., Ghosh, M.K., Chattopadhyay, D.J. and Chatterjee, I.B. 2000. Vitamin C prevents cigarette smoke-induced oxidative damage *in vivo.* Free Radic. Biol. Med. 29: 115-124.

Panda, K., Chattopadhyay, R., Ghosh, M.K. Chattopadhyay, D.J. and Chatterjee, I.B. 2001. Cigarette smoke-induced protein oxidation and proteolysis is exclusively caused by its tar phase: Prevention by Vitamin C. Toxicol. Lett. 123: 21-23.

Pannala, A.S., Rice-Evans, C.A., Halliwell, B. and Singh, S. 1997. Inhibition of peroxynitrite-mediated tyrosine nitration by catechin polyphenols. Biochem. Biophys. Res. Commun. 232: 164-168.

Pianzza, M.L., Sellers, E.M. and Tyndale, R.F. 1998. Nicotine metabolism defect reduces smoking. Nature 393:750.

Pryor, W.A., Deoley, M.M. and Chrch, D.F. 1986. The inactivation of α_1-proteinase inhibitor by gas-phase cigarette smoke: Protection by antioxidants and reducing species. Chem. Biol. Interact. 57: 271-183.

Pryor, W.A., Prier, D.G. and Church, D.F. 1983. Electron spin resonance study of mainstream and side stream cigarette smoke: Nature of the free radicals in gas phase smoke and cigarette tar. Environ. Health Perspect. 47: 345-355.

Reznick, A.Z., Cross, C.E., Hu, M.L., Suzuki, Y.J., Khwaja S. Safadi, A., Mottchnick, P.A., Packer, L. and Halliwell, B. 1992. Modification of plasma proteins by cigarette smoke as measured by protein carbonyl formation. Biochem. J. 286: 607-611.

Sano, M., Takahashi, Y., Yoshino, K., Shimoi, K., Nakamura, Y., Tomita, I., Oguni, I. and Konomoto, H. 1995. Effect of tea (*Camellia sinensis* L.) on lipid peroxidation in rat liver and kidney: a comparison of green and black tea feeding. Biol. Pharmaceutical Bull. 18: 1006-1008.

Schraufstatter, I.U., Revak, S.D. and Cochrane, C.G. 1984. Protease and oxidants in experimental pulmonary inflammatory injury. J. Clin. Invest. 73: 1175-1184.

Serafini, M., Ghisellii, A. and Ferro-Luzzi, A. 1996. *In vivo* antioxidant effect of green and black tea in man. Eur. J. Clin. Nutr. 50: 28-32.

Shah, P.K. and Helfant, R.H. 1998. Smoking and coronary heart. Chest. 94:449-452.

Shalini, V.K., Luthra, M., Srinivas, L., Rao, S.H. and Balasubhramaniam, D. 1994. Oxidative damage to the eye lens caused by cigarette smoke and fuel smoke condensate. Indian J. Biochem. Biophys. 314: 261-266.

Sherman, P.K. 1991. Health effect of cigarette smoking. Clin. Chest Med. 12: 643.

Shim, J., Kang, M., Kim, Y., Roh, J., Roberts, C. and Lee, I. 1995. Chemo-preventive effect of green tea among smokers. Cancer Epidermis. Biomarkers Prev. 4: 387-391.

Subhramanian, N., Venkatesh, P., Ganguli, S. and V.P. Sinkar, 1999. J. Agric. Food Chem. 47: 2571-2578.

Ueyama, K., Yokode, M., Arai, H., Nagano, Y., Li, Z. X, Cho, M. and Kita, T. 1998. Cholesterol efflux effect of high-density lipoproteins is impaired by whole phase cigarette smoke extracts through lipid peroxidation. Free Radic. Biol. Med. 24(1): 182-90.

U.S. Surgeon General's Report. Department of Health Education and Welfare, Department of Health and Human Services, USA, 1985.

Veyssier, B.C. 1997. Tobacco smoking and cardiovascular risk. Rev. Med. Interne. 189: 702-708.

Weisburger, J.H. 1997. Tea and health: a historical perspective. Cancer Lett. 114: 315-317.

Wei, H., Zhang, X., Zhao, J.F., Wanh, Z.Y., Bickers, D. and Lebwohl, M. 1999. Scavenging of hydrogen peroxide and inhibition of ultraviolet light-induced oxidative DNA damage by aqueous extracts from green and black tea. Free Radic. Biol. Med. 26: 1427-1435.

Weisburger, J.H. 1999. Tea and health: the underlying mechanisms. Proc. Soc. Exp. Biol. Med. 220: 271-275.

Wiseman, S.A., Balentine, D.A. and Frie, B. 1997. Antioxidant in tea. Crit. Rev. Food Sci. Nutr. 37: 705-718.

Yang, C.S. and Wang, Z.Y. 1993. Tea and cancer. J. Natl. Cancer Inst. 85: 1028-1049.

Yang, C.S. 1997. Inhibition of carcinogenesis by tea. Nature 389: 134.

Yokode, M., Ueyama, K. and Arai, N.H. 1996. Modification of high- and low-density lipoproteins by cigarette smoke oxidants. Ann. N.Y. Acad. Sci. 786: 245-251.

Zhang, A., Zhu, Q.Y., Luk, Y.S., Ho, K.Y., Fung, K.P. and Chen, Z.Y. 1997. Inhibitory effects of Jasmine green tea epicatechins isomers on free radical-induced lysis of red blood cells. Life Sci. 61: 383-394.

[illegible]

Chapter 9

Cancer Chemoprevention by Tea Polyphenols

Rajesh Krishnan, Rachana Patel,
Asha G. Ramchandani and Girish B. Maru
Tobacco Carcinogenesis Group,
Advanced Centre for Treatment, Research and
Education in Cancer (ACTREC), Tata Memorial Centre,
Kharghar, Navi-Mumbai-410 208, India.
e-mail: gmaru@actrec.res.in

INTRODUCTION

Majority of human cancers are attributed to environmental factors including pollutants in the air and water, work place exposure, radiation and personal habits such as smoking/tobacco chewing and dietary patterns. Further identification of specific causative factors and evaluation of their relative importance have proved to be rather difficult since a majority of cancers result from complex interactions between environmental and host factors. Efforts to eliminate known human carcinogens from the environment and current cancer treatment approaches have met with limited success. Based on the experience with some infectious diseases and the recent progress in cardiology, prevention of diseases appears to be one of the attractive approaches. Similarly, based on experimental and epidemiological evidence, chemoprevention or dietary intervention especially employing plant-derived antioxidants, is receiving increasing attention. A number of plant-derived antioxidants have shown anti-initiating and/or anti-promoting activities in experimental systems (Wattenberg, 1987; Cao et al., 1996; Weisburger and Chung, 2002; Frei and Higdon, 2003). One such candidate, which needs further studies, is tea (*Camellia sinensis*), which is one of the most popular beverages worldwide. Presented below is the review and current status with reference to cancer prevention by tea polyphenols and some information from our recent results.

Based on the manufacturing process (Fig. 9.1; also see xxi and pp 1-3 this volume) employed (Harbowy and Balentine, 1997), tea can be classified as:

(i) Green tea (≈20-22% world tea consumption)
(ii) Oolong tea (≈ 2-3% world tea consumption)
(iii) Black tea (≈75% -78% world tea consumption)

White tea (Buds or young leaves)	Steamed (oxidase inactivation) → Dried
Green tea (mature leaves)	Withered → Steamed or Panfried (oxidase inactivation) → Dried
Oolong tea (mature leaves)	Withered → Oxidised briefly → Panfried → Dried
Black tea (mature leaves)	Withered → Rolled → Oxidised fully → Fried → Dried

Fig. 9.1 Methods of manufacturing different types of teas

BIOLOGICAL EFFECTS

Different types of tea vary from each other in respect of the nature and amounts of flavonoids present and many of the biological activities of tea appear to be related to its flavonoid content. Flavanols and flavonols are groups of related, naturally occurring chemicals in plants. A number of components in these classes act as antioxidants which may protect against oxidative stress. Catechins are flavanols predominating in tea leaves viz. (–) epicatechin (EC), (–) epicatechin gallate (ECG), (–) epigallocatechin (EGC) and (–) epigallocatechin gallate (EGCG). It is mainly to preserve the catechins that green tea is typically panfried to inactivate polyphenol oxidase (PPO) as a result 60-80% of total flavonoids in green tea are catechin monomers (Figs. 9.2a and b).

The biological effects of green tea and its individual polyphenols have been reviewed extensively (Lodovici et al., 2000; Frei and Higdon, 2003). Various studies employing different *in vivo* and *in vitro* models suggest that

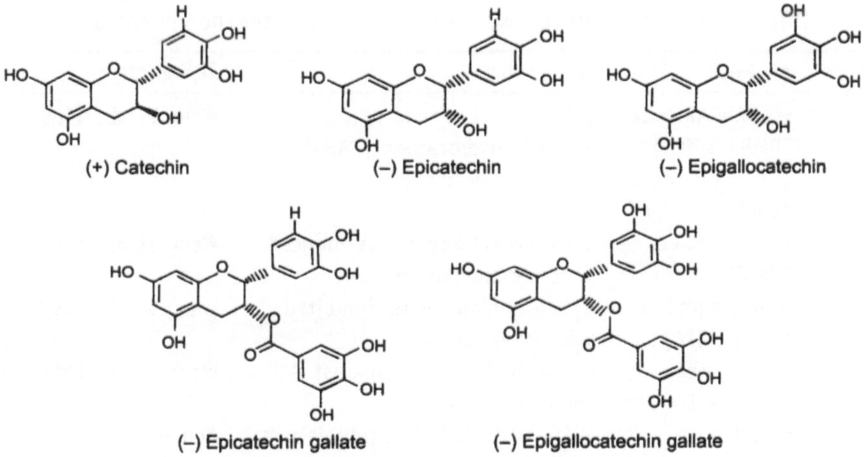

Fig. 9.2a Structure of free catechin monomers

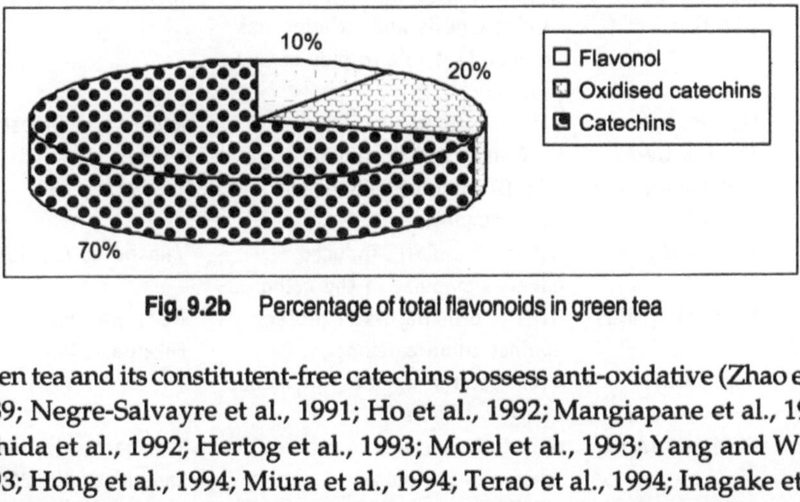

Fig. 9.2b Percentage of total flavonoids in green tea

green tea and its constitutent-free catechins possess anti-oxidative (Zhao et al. 1989; Negre-Salvayre et al., 1991; Ho et al., 1992; Mangiapane et al., 1992; Uchida et al., 1992; Hertog et al., 1993; Morel et al., 1993; Yang and Wang 1993; Hong et al., 1994; Miura et al., 1994; Terao et al., 1994; Inagake et al., 1995; Jovanovich et al., 1995; Yen and Chen, 1995; Cao et al., 1996; Fourneau et al., 1996; Guo et al., 1996; Knekt et al., 1996; Nanjo et al., 1996; Ishikawa et al., 1997; Katiyar and Mukhtar, 1997a, Luo et al., 1997; Pannala et al., 1997; Yu et al., 1997; Gardner et al., 1998; Mazur et al., 1998; Trevisanato and Kim 2000; Reimersma et al., 2001; Frei and Higdon, 2003; Yanagimoto et al., 2003; Negishi et al., 2004), anti-inflammatory (Katiyar et al., 1993a; 1995, Chan et al., 1995; Knekt et al., 1996; Lin and Liang, 2000, Hong et al., 2001; Reimersma et al., 2001), anti-mutagenic (Nagabhushan et al., 1991; Mukhtar et al., 1992; Bu-Abbas et al., 1994; Yen and Chen, 1995, Constable et al., 1996; Catterall et al., 2003b) as well as anti-carcinogenic activities (Table 9.1).

Table 9.1 Chemopreventive effect(s) of green tea/polyphenols

Sr. No.	*Model(s)*	*Conclusions*	*References*
1.	(i) SENCAR mice [DMBA + TPA] (ii) Balb-C mice [3-MC]	GTPs protect against polycyclic aromatic hydrocarbon [PAHs]-induced skin tumors	Wang et al., 1989
2.	Hairless mice [UV-B]	GTPs protect against UV-induced photocarcinogenesis	Wang et al., 1991
3.	SKH-1 mice [UV-B + TPA]	GTE in drinking water inhibited skin carcinogenesis	Wang et al., 1992d
4.	SKH-1 mice [DMBA + UV-B]	GTE or DGTE protect against skin carcinogenesis	Wang et al., 1994
5.	SKH-1 mice [UV-B]	EGCG, GTPs, GTE, BTE inhibit skin carcinogenesis	Agarwal et al., 1994, Gensler et al., 1996, Yamane et al., 1996
6.	SKH-1 mice [UV-B]	GTE inhibits UV-induced carcinogenesis and caffeine has an important role to play in this process	Huang et al., 1997
7.	Hairless mice [UV-B & UV-A]	GTE protects against UV-B and UV-A induced skin tumors	Record and Dreosti 1998a; b
8.	SKH-1 mice [UV-B]	GTE, BTE inhibit epidermal carcinogenesis	Yen et al., 1997
9.	Rat [ENNG]	GTP inhibits ENNG-induced adenocarcinomas in the colon	Yamane et al., 1996
10.	F344 rats [MNU]	GTPs in drinking water protect against colon carcinogenesis	Narisawa and Fakaura, 1993
11.	Mice [DMH]	GTE, catechin decrease incidence of colon tumors	Yin et al., 1994
12.	F344 rats [AOM]	GTE, GTPs decrease number of adenomas in colon	Yamane et al., 1991; Caderni et al., 2000
13.	A/J mice [B(a)P , NDEA]	WEGT inhibits B(a)P induced forestomach tumors and NDEA induced lung tumors	Wang et al., 1992a; Xu et al., 1992
14.	A/J mice [DEN,NNK]	GTE and its polyphenol(s) inhibit TSNA-induced lung tumorigenesis	Wang et al., 1992b
15.	A/J mice [NNK]	GTE and DGTE inhibit incidence and multiplicity of NNK-induced lung tumors	Shi et al., 1994
16.	SD rats [DMBA]	GTE catechin protects against mammary gland carcinogenesis	Hirose et al., 1994
17.	SD rats	GTE decreases the incidence and	Wang et al., 1995

Contd.

Table 9.1 contd.

Sr. No.	*Model(s)*	*Conclusions*	*References*
	[NMBzA]	multiplicity of NMBzA-induced esophageal tumors	
18.	C3H/HeN Crj mice	EGCG inhibits spontaneous hepatoma formation	Nishida et al., 1994
19.	CD-1 mice [DMBA, B(a)P+TPA]	GTP decreases tumor multiplicity and incidence and also inhibited TPA-induced hyperplasia, ODC activation and H_2O_2 formation	Huang et al., 1997
20.	SENCAR mice [DMBA + TPA]	GTPs show anti-promoting activity	Agarwal et al., 1992; Katiyar et al., 1992
21.	Rat multi-organ carcinogenesis model [NDEA, MNU, AOM, etc.]	GT catechin in diet inhibits initiation and post-initiation phases of tumorigenesis	Hirose et al., 1993
22.	SENCAR mice [DMBA+TPA – Stage I] [DMBA+ mezerein– Stage II]	GTPs inhibit stage I as well as stage II of tumor promotion	Katiyar et al., 1993d
23.	WB-F344 rat liver epithelial cells [DDT, TPA, dieldrin]	GTPs enhance the gap junctional intercellular communication between cells, which is reduced by DDT, TPA or dieldrin, suggesting anti-promoting activity	Seigler and Ruch, 1993
24.	JB6 mouse epidermal cell line	EGCG and TFs inhibit EGF- or TPA-induced JNK and AP-1 activity, thereby inhibiting neoplastic transformation	Dong et al., 1997
25.	SENCAR mice [4-NQO, BPO]	GTP possess significant preventive effects against tumor progression induced by genotoxic agents [NQO] or free radical generating compounds [BPO]	Katiyar et al., 1993a
26.	(i) A/J mice [Complete carcinogenesis] [NDEA, B(a)P] in both studies	(i) GTP, WEGT provide significant protection at all stages of carcinogenesis (lung & forestomach)	Katiyar et al., 1993b,e
	(ii) SENCAR mice [DMBA-TPA]	(ii) GTPs protect against malignant conversion of chemically-induced	Katiyar et al., 1993c

Contd.

Table 9.1 Contd.

Sr. No.	*Model(s)*	*Conclusions*	*References*
		benign skin papillomas to squamous cell carcinomas	
27.	A/J mice	GTE, BTE inhibit spontaneous formation of lung tumors and rhabdomyosarcomas	Landau et al., 1997
28.	(i) Mice (ii) B16 melanoma cells	EGCG inhibits lung metastasis by lung carcinoma cells	Taruguchi et al., 1992
29.	SKH-1 mice [UV-B]	GTE inhibits growth of established skin papillomas	Wang et al., 1992c
30.	Different *in vitro* model systems	GTPs, free catechins, BTPs, TFs show anti-mutagenic, anti-proliferative and anti-neoplastic activities	Steele et al., 2000
31.	Human bronchial epithelial cell lines (SV-40 immortalized 33 BES cells, H-ras transformed 21 BES cells)	Free catechins, TFs decrease c-jun protein phosphorylation, which possibly lowers AP-I activity and hence inhibits growth of cells.	Yang et al., 2000
32.	HTC rat hepatoma cells DS19 mouse erythroleukeamia cells	EGCG, GTE inhibit DNA synthesis and cell proliferation	Lea et al., 1993
33.	Rat liver microsomes	Free catechins, GTPs inhibit the AHH, ECOD and EROD activities of CYP450s and thus inhibit CYP450-mediated binding of B(a)P, DMBA to DNA	Wang et al., 1988
34.	F344 rats	WEGT induces the MROD & PROD activities of CYP450 isoforms and levels of 1A2 & 4A1	Bu-Abbas et. al., 1994
35.	F344 rats	(i) GTE increases levels of 1A1, 1A2 and 2B1, but no change in 2E1 and 3A4. (ii) Increase in UGT, but not GST GT modulates phase I and phase II enzymes	Sohn et al., 1994
36.	Mice	GTE inhibits UV-B induced DNA	Chatterjee et al.,

Contd.

Table 9.1 Contd.

Sr. No.	*Model(s)*	*Conclusions*	*References*
		lesions as detected by ^{32}P post-labelling analysis	1996
37.	F344 rats [IQ]	GTE effectively reduces the incidence of aberrant crypts and DNA adducts in rat colon	Xu et al., 1992
38.	F344 rats [PhIP]	GTE inhibits PhIP-DNA adducts in small intestine, colon, liver and mammary gland epithelial cells	Schut and Yao, 2000

AOM=Azoxymethane, B(a)P=Benzo(a)pyrene, BPO=Benzoyl peroxide, DDT=Dichlorodiphenyl trichloroethane, DGTE=Decaffeinated green tea extract, DMBA=7,12-Dimethylbenz(a) anthracene, DMH=1,2-Dimethylhydrazine, ENNG=N-ethyl-N'-nitro-N-nitrosoguanidine, GTE=Green tea extract, GTPs=Green tea polyphenols, MC=3-Methylcholantherene, MNU=N-methyl-N-nitrosourea, NDEA=N-Nitrosodiethylamine, NMBzA=N-Nitrosomethylbenzylamine, NNK=4-(methylinitrosamino)-1-(3-pyridyl)-1-butanone, NQO=4-nitroquinoline 1-oxide, PhIP=2-amino-1-methyl-6-phenylimidazo [4,5-b] pyridine, TPA=12-*O*-tetradecanoylphorbol-13-acetate, WEGT=Water extract of green tea.

To summarize, green tea and its polyphenols have been shown to prevent carcinogenesis at various sites such as skin, forestomach, liver, lung, oesophagus, colon, small intestine and mammary glands employing different models. All the above-mentioned protective effects of green tea have been attributed to the catechin and gallocatechin monomers.

Studies have also been carried out on humans with the aim to relate green tea consumption with cancer risk (Table 9.2).

By and large, most of these studies have been inconclusive, partly because they were not specifically designed to assess the risk of cancer in conjunction with green tea consumption.

CHEMOPREVENTION

As mentioned above, the chemopreventive efficacy of green tea has been attributed to the presence of free catechins. However, during the manufacture of black tea, there is a marked reduction in the amount of free catechin monomers due to oxidative polymerization, catalyzed by the enzyme polyphenol oxidase (PPO) (Subramanian et al., 1999), to form a new set of compound(s), viz., oligomeric theaflavins (TFs) and polymeric thearubigins (TRs), which constitute almost 13% and 47%, respectively, of total black tea polyphenols (Figs. 9.3a and 9.3b).

Table 9.2 Epidemiologic studies on green tea consumption and cancer

Sr. No.	*Cancer*	*Country*	*Type of study*	*Subjects (follow-up period)*	*Association*	*References*
1.	All sites	Japan	Cohort	8,522 men and women 384 cancer cases (9 yrs)	Inverse, significant for women only	Imai et. al. 1997
2.	Esophageal	China	Case-control	1,552 controls 902 cancer cases	Inverse, significant for women only; inconclusive for men	Gao et. al. 1994
3.	Gastric	Japan	Cohort	11,902 men 14,409 women 296 cases in men 123 cases in women (8 yrs)	Non-significant	Tsubono et. al. 2001
4.	Pancreatic	China	Case-control	1,552 controls 451 cancer cases	Inverse, but not conclusive	Ji et. al. 1997
5.	Bladder	Japan	Cohort	38,450 atomic bomb survivors 114 cases (12 – 14 yrs)	Non-significant	Nagano et. al. 2000

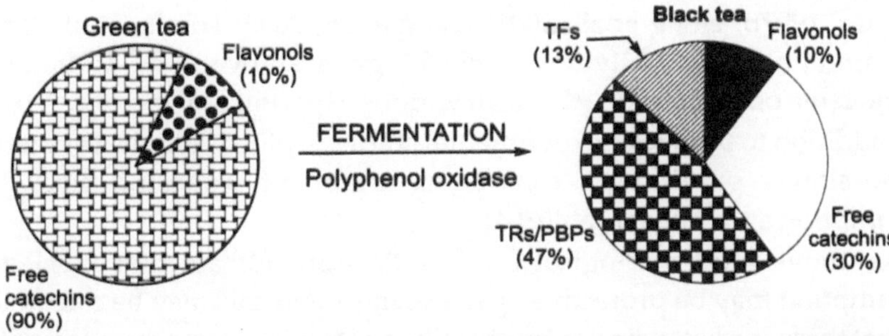

Fig. 9.3a Oxidative polymerization of free catechins during black tea manufacturing

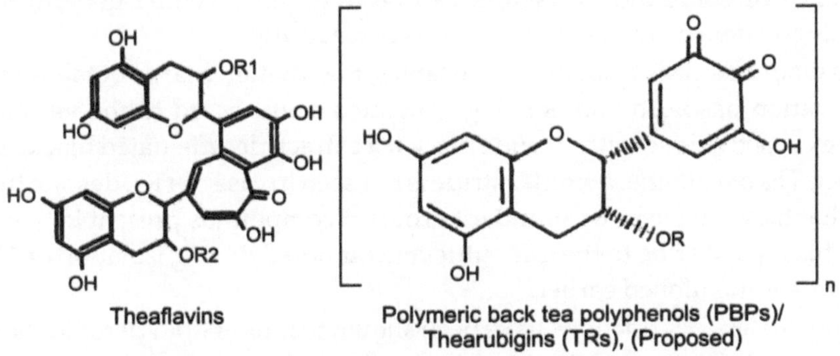

Fig. 9.3b Structures of characteristic polyphenols in black tea

Since the discovery by Roberts et al., (1957) of TFs and TRs in black tea, a lot of information has been generated with regard to the chemical structures and biological effects of TFs (Roberts and Smith 1961; Bryce et al., 1970; Coxon et al., 1970; Collier and Mallows, 1971; Hoefler and Coggon, 1976; Robertson and Bendal, 1983; Kiehne and Engelhardt, 1996; Lee et al., 2000), but TRs, due to their high matrix reactivity, have been very difficult to isolate and characterize till date, and very little information is available about their biological activity (Wedzicha and Donovan, 1989; Finger et al., 1992; Haslam, 2003).

Several studies undertaken employing crude black tea extract, decaffeinated black tea and TFs have shown that black tea possesses anti-oxidative (Shiraki et al., 1994; Yoshino et al., 1994; Miura et al., 1995; Salah et. al 1995; Yen and Chen, 1995; Miller et al., 1996; Wei et al., 1999; Zhang et al., 2000a; Sarkar and Bhaduri, 2001; Saha and Das, 2002; Yanagimoto et. al 2003; Negishi et al., 2004), anti-mutagenic (Nagabhushan et al., 1991; Yen and Chen, 1995; Bu-Abbas et al., 1996; Constable et al., 1996; Gupta et al., 2001; 2002, Dhawan et al., 2002; Catterall et al., 2003a; b), as well as anti-inflammatory (Maity et al., 1995; 1998, Clarke et al., 1997; Katiyar and

Mukhtar, 1997b; Hong et al., 2001; Liang et al., 2002; Tomita et al., 2002; Wolfram et al., 2002) activities. Table 9.3 gives an overview of the anti-carcinogenic potential of black tea, in various experimental models.

In addition to the above experimental studies, epidemiological evidence is also slowly providing a convincing argument regarding black tea consumption and cancer (Table 9.4).

The above epidemiological studies do not indicate that black tea consumption may be protective against cancer and this may be due to the fact that these studies were not specifically designed to assess the effect of tea on the risk of cancer. There is an urgent need for such studies so that the influence of confounding factors can be minimized, while generating meaningful data on the protective effects of tea, if any.

As would be clear from the above tables, that though a large database of information has been and is being generated with regard to the various biological activities of either crude black tea extract, decaffeinated black tea extract, TFs or caffeine, a similar stride is not seen in case of TRs, despite the fact that they are the most abundant group of compounds present in black tea. This is partly due to the difficulties encountered during isolation of TR fractions as mentioned earlier.

Nonetheless, TR fractions have been shown to inhibit lipid peroxidation in rat liver homogenates induced by tert-butyl hydroperoxide (Yoshino et al., 1994), decrease the occurrence of chromosomal aberrations and sister chromatid exchanges in dimethylbenz(a)anthracene (DMBA) or cyclophosphamide (CP) treated Swiss albino mice (Gupta et al., 2001), inhibit 8-OHdG formation in DNA from colon mucosa cells in rats treated with dimethylhydrazine (DMH) and also modulate intestinal CYP450 and conjugase activities in rats (Lu et al., 2000; Catterall et al., 2003c). Apart from the above studies, not much information is available from the existing literature with regard to the biological activities of TRs. Hence, attention has been focussed on isolating and studying the biological effect(s) of black tea TRs, employing various *in vitro* and *in vivo* model systems.

Roberts et al., (1957) referred to this ill-defined group of substances as thearubigins (TRs) which are water-soluble, acidic and display an ill-defined chromatographic behaviour. TRs were partially fractionated by Roberts and Smith (1961) and the compounds in black tea liquor co-extracting into ethyl acetate together with TFs were designated—SI thearubigins. Those remaining in the aqueous phase were designated—SII and SI(a) TRs, the latter having enhanced solubility in diethyl ether.

Brown et al., (1969) separated thearubigins from black tea liquors by solvent extraction using ethyl acetate, n-butyl alcohol and acidified n-butyl

Table 9.3 Chemopreventive effect(s) of black tea/polyphenols

Sr. No.	*Model(s)*	*Compounds tested*	*Conclusions*	*References*
1.	C3H mice [NDEA]	BTE	BT inhibits NDEA-induced lung and liver tumors in a dose-dependent manner	Cao et al., 1996
2.	Hairless mice [UV-B & UV-A]	BTE	BT protects against UV-B and UV-A induced skin tumors	Record and Dreosti, 1998 a; b
3.	SKH-1 mice [UVB]	SBTE, BTF1, BTF2	BT and its components abrogate UVB-induced erythema and associated early events	Zhao et al., 1999
4.	Rats [IQ and DMBA]	BTE/BTE+milk	Tea/Tea with milk reduce aberrant crypt formation in IQ-treated rat, only during initiation, and not later	Weisburger et al., 1997
5.	A/J mice [NMBzA]]	TFs & EGCG	Only TFs reduced multiplicity of esophageal tumors	Morse et al., 1997
6.	A/J mice [NNK]	BTP, TFs	BTP and TF mixture inhibit NNK-induced lung adenomas and cell proliferation	Morse et al., 1989, Yang et al., 1997 a; b
7.	Mice [DMBA + TPA]	BTPs	BTPs show both anti-initiating and anti-promoting activities	Javed et al., 1998
8.	SKH-1 mice [UV B, complete carcinogenesis]	GTE, BTE, DGTE, DBTE, caffeine	Caffeine is a biologically important constituent of tea	Huang et al., 1997
9.	A/J mice	GTE, BTE	Inhibit spontaneous formation of lung tumors and rhabdomyosarcomas	Landau et al., 1997
10.	F344 rats [DMH]	Black tea complex polyphenols (TFu fraction)	Complex BT polyphenols significantly inhibit DMH-induced oxidative DNA damage in colon mucosa of rats	Lodovici et al., 2000
11.	SKH-1 mice [UVB]	BTE	BT induces apoptosis and decreases mitosis, which as seen by decrease in BrdU incorporation in DNA and regression of tumor	Lu et al., 1997

Contd.

Table 9.3 Contd.

Sr. No.	*Model(s)*	*Compounds tested*	*Conclusions*	*References*
12.	Guinea pigs (partially ascorbate deficient) [CS]	BT infusion	BT and its polyphenols reduce cigarette smoke-induced modifications of proteins and degradation of microsomal proteins	Misra et al., 2003
13.	Swiss albino mice [DEN]	BTE	BT extract inhibits DEN-induced pulmonary adenocarcinomas	Shukla and Taneja, 2002
14.	Human bronchial epithelial cell lines (SV-40 immortalized 33 BES cells, H-ras transformed 21 BES cells)	Free catechins, TFs	Decrease c-jun protein phosphorylation, which possibly lowers AP-I activity and hence inhibits growth of cells.	Yang et al., 2000
15.	Mouse epidermal JB6 cell line with mutant H-ras gene [also called as 30.7b Ras 12 cells]	Catechins, TFs	(i) Both inhibit cell growth and AP-1 activity (ii) Both inhibit phosphorylation of P44/42 (ERK1/2) and c-jun without affecting levels of phosphorylated JNK. (iii) TFdiG inhibits phosphorylation of p38, but EGCG does not. (iv) EGCG decreases levels of c-jun, while TFdiG decreases levels of fra-1 Tea polyphenols inhibit AP-1 activity and MAP kinase pathway, which contribute to growth inhibition, however different mechanisms may be involved in inhibition by free catechins and TFs	Chung et al., 1999
16.	JB6 mouse epidermal cell line	EGCG, TFs	Inhibit EGF- or TPA-induced JNK and AP-1 activity, thereby inhibiting neoplastic transformation	Dong et al., 1997

Contd.

Table 9.3 Contd.

Sr. No.	*Model(s)*	*Compounds tested*	*Conclusions*	*References*
17.	Different *in vitro* model systems	GTPs, free catechins, BTPs, TFs	Show anti-mutagenic, anti-proliferative and anti-neoplastic activities	Steele et al., 2000
18.	(i) HPAC (Pancreatic adenocarcinoma) (ii) LnCaP (prostate tumor) cell lines (human-derived)	GTP, BTP, TFs and free catechins	GTP & BTP modulate genes which are known to play a role in the carcinogenesis process (K-ras & mdr-1 gene)	Lyn-Cook et al., 1999
19.	(i) SV40 transformed W138 VA cells (ii) Caco-2 colon cancer cells	TFs	TFs inhibit growth and induce apoptosis in W138 VA cells. They also inhibit Cox-2 gene expression at both the mRNA and protein levels	Lu et al., 2000
20.	Mouse JB6 epidermal cell line [UV-B]	TFs	TFs inhibit UV-B induced AP-1 activation by inhibiting the activation of ERKs and JNK.	Nomura et al., 2000
21.	(i) Mouse [TPA] (ii) NIH3T3 cells [TPA]	TF-1, TF-2, TF-3 and EGCG	(i) Inhibit TPA-induced ear edema and ODC activity in skin epidermis and NIH3T3 cells.	Liang et al., 2002
22.	Prostate cancer cell lines DU145 (Androgen-unresponsive) LNCaP (Androgen-responsive)	TFs & EGCG	(i) Inhibit PI-3K/PKB and phosphorylation of Akt (ii) Induce ERK 1/2 in both cell lines EGCG and TFs modulate constitutive activation of PI-3K/Akt and Erk 1/2 pathways.	Siddiqui et al., 2004
23.	Four human cancer cell lines: (i) H 661 (lung tumor cell line)	TF, GTP, DGTE, GTE	Tea polyphenols induce apoptosis and therefore contribute to growth inhibitory activities TF > GTP > DGT > GTE	Yang et al., 1998 a; b

Contd.

Table 9.3 Contd.

Sr. No.	*Model(s)*	*Compounds tested*	*Conclusions*	*References*
	(ii) H 1299 (lung tumor cell line) (iii) H 441 (lung cancer cell line) (iv) HT-29 (colon cancer cell line)			
24.	Swiss albino mice [DMBA or CP]	TF & TR & BTE	Significantly decrease both CA & SCE induced by either DMBA or CP	Gupta et al., 2001
25.	SD rats [AOM]	BT infusion	BT infusion significantly inhibits cell proliferation and induces apoptosis thereby reducing the number of aberrant crypts. BT protects against colon carcinogenesis	Sengupta et al., 2002
26.	Murine myeloid leukaemia WEHI-3BJCS cells	TF-1, TF-2A, TF-2B, TF-3 and EGCG	Dose-dependent anti-proliferative and anti-apoptotic effects. TF-2B & TF-3 most potent in reducing both the *in vitro* clonogenicity and *in vivo* tumorigenicity of WEHI-3BJCS cells	Lung et al., 2004
27.	Rat ascites hepatoma AH109A cells B16 melanoma cells	Free catechins, TFs	Free catechins, their metabolites, TFs inhibit ROS generated by hypoxanthine-xanthine oxidase system, thereby preventing ROS mediated adhesion and invasion of underlying monolayer.	Zhang et al., 2000 a
28.	N-hydroxylation of PhIP and ABP, using rat and human liver microsomes	Catechins, TFs, etc.	TFs were judged to be primary inhibitory components in tea, and show most potent effect.	Hammons et al., 1999

Contd.

Table 9.3 Contd.

Sr. No.	*Model(s)*	*Compounds tested*	*Conclusions*	*References*
29.	SD rats	Aqueous extract of GT or BT, GTPs, BTPs	GT, BT and their polyphenols inhibit formation of N-acetoxy PhIP, the ultimate carcinogenic form of PhIP	Lin et al., 2003
30.	Mouse skin model [DMBA]	TFs	TFs inhibit cell proliferation and lipid peroxidation; TFs induce apoptosis and also significantly elevate levels of detoxification enzymes (GST, GPx, SOD and CAT)	Saha and Das, 2002
31.	Rats	TFs, TFu	TFs and theafulvins (TFu) inhibit the activity and levels of CYP1A1, 2E1 and 3A; but not that of CYP 4A1. TFs do not have any effect on Phase II enzymes, while TFus inhibit glucuronosyl transferase activity, in intestine. No changes observed in liver of rats.	Catterall et al., 2003c
32.	SD rats [AOM]	BT infusion	BT infusion significantly elevates the levels of GST & GPx and inhibits lipid peroxidation. This causes a decrease in ACFs during initiation phase.	Sengupta et al., 2003
33.	F344 rats [AOM]	BTE	Black tea inhibits formation of O^6methyl & N^7methyl products in the DNA of rat colon cells	Chen et al., 1998
34.	Calf thymus DNA	BTPs	BTPs inhibit formation of H_2O_2 induced 8-OHdG in calf thymus DNA	Wei et al., 1999

ABP=2-Aminobiphenyl, AOM=Azoxymethane, BITC = Benzyl isothiocyanate, BTE = Black tea extract, BTF=Black tea fraction, BTPs = Black tea polyphenols, DBTE = Decaffeinated black tea extract, DEN = Diethylnitrosamine, DGTE = Decaffeinated green tea extract, GTE = Green tea extract, I-3-C = Indole-3-carbinol, IQ = 2-amino-3-methylimidazo[4,5-f]quinoline, NNK= 4-(methylnitrosamino)-1-(3-pyridyl)-1-butanone, ODC=Ornithine decarboxylase, PhIP=2-amino-1-methyl-6-phenylimidazo[4,5-b]pyridine, SBTE=Standardized black tea extract, TF(s) = Theaflavin(s), TFu = Theafulvins, TPA= 12-*O*-tetradecanoyl phorbol-13-acetate.

Table 9.4 Epidemiological studies on black tea and cancer

Sr. No.	*Cancer*	*Country*	*Type of study*	*Subjects (follow-up period)*	*Association*	*References*
1.	All sites	Netherlands	Cohort	738 men 75 cases (5 yrs)	Non-significant	Hertog et al., 1993
2.	Gastric	Netherlands	Cohort	58,279 men 62,573 women 160 cases in men 40 cases in women (4.3 yrs)	Non-significant	Goldbohm et al., 1996
3.	Gastric	U.S., Hawaii	Cohort	5,233 men 6,297 women 64 cases in men 44 cases in women (14.8 yrs)	Non-significant	Galanis et al., 1998
4.	Colorectal	Netherlands	Cohort	58,279 men 62,573 women 650 cases (4.3 yrs)	Non-significant	Goldbohm et al., 1996
5.	Rectal	Finland	Case-control	29,133 male smokers 111 cases	Non-significant	Hartman et al., 1998
6.	Pancreatic	U.S.	Case-control	13,979 men and women 65 cases (9 yrs)	Inverse	Shibata et al., 1994
7.	Pancreatic and bladder	U.S.	Cohort	47,794 men 130 cases (12 yrs) 88,799 women 158 cases (16 yrs)	Non-significant for men, women and pooled sample	Michaud et al., 1999
8.	Lung	Netherlands	Cohort	58,279 men 62,573 women 764 cases (4.3 yrs)	Non-significant	Goldbohm et al., 1996
9.	Bladder	Japan	Cohort	38,450 atomic bomb survivors 114 cases (12–14 yrs)	Non-significant	Nagano et al., 2000
10.	Prostate	Canada	Cohort	3400 men	Non-significant	Ellison, 2000
11	Breast	Netherlands	Cohort	62,573 women 507 cases (4.3 yrs)	Non-significant	Goldbohm et al., 1996

alcohol. TR-1 was analogous to SI of Roberts and Smith (1961) while TR-2 to TR-5 were further separations of SII and SI(a) fractions. Each fraction was hydrolyzed and when subjected to 2-D paper chromatography revealed pink spots, characteristic of proanthocyanidins. They, therefore, proposed that all TRs had a general structure based on that of polymeric proanthocyanidins, which has been one of the most contentious issues of thearubigin research to date (Cattel and Nursten, 1976; Haslam, 2003).

Several groups have tried separation of TRs employing Sephadex LH-20 chromatography (Crispin et al., 1968, Bailey et al., 1992; Finger et al., 1992; Harbowy and Balentine, 1997) and reverse phase HPLC (Hoefler and Coggon, 1976; Robertson and Bendal, 1983; Wedzicha and Donovan, 1989; Opie et al., 1990; Harbowy and Balentine, 1997) without much progress. When subjected to HPLC these compounds show a propensity to elute as an "unresolved Gaussian hump (Robertson and Bendal, 1983; Opie et al., 1990). Wedzicha and Donovan (1989) suggested that due to high affinities between polyphenolics and silica-based HPLC stationary phases, recoveries were poor and have employed counter-current chromatography for separation of TRs. Relatively modern methods such as liquid chromatography-mass spectrometry (LC-MS), high speed counter-current chromatography (HSCCC) and centrifugal precipitation chromatography (CPC) are also being employed for preparative separation of TRs (Kiehne and Engelhardt, 1996; Degenhardt et al., 2000; 2001).

Despite all these advances, the fact remains that though TRs are the most abundant products of oxidative polymerization of free catechins during black tea manufacture, they are the substances whose chemistry is least understood (Harbowy and Balentine, 1997; Haslam, 2003).

Taking the above aspect into consideration, TRs were extracted employing the method of Brown et al., (1969). TR fractions (henceforth referred to as polymeric black tea polyphenol fractions, PBPs), thus isolated, were checked for their color, pH, TLC mobility, purity/absence of other known biologically active components, such as caffeine, free catechins or theaflavins, proanthocyanidin nature, $FeCl_3$ reactivity, UV absorption (Fig. 9.4), FT-IR (Fig. 9.5) and NMR spectra (Fig. 9.6) (Table 9.5).

Isolated PBPs were then tested for their biological activity, individually or as a mixture, employing *in vitro* and *in vivo* model systems.

An important approach to chemoprevention is to block the DNA damage caused by carcinogenic insult which is referred to as initiating stage of carcinogenesis. Xenobiotic chemicals, including carcinogens, which gain entry into the cell, are acted upon by xenobiotic metabolizing enzymes

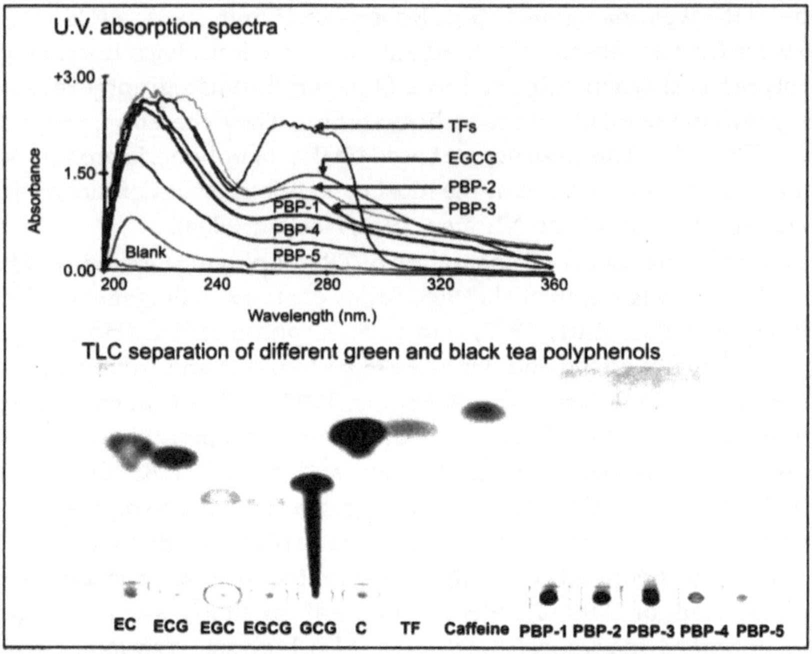

Fig. 9.4 UV Spectra and TLC mobility of green and black tea polyphenols
EC=(–)-Epicatechin
ECG=(–)-Epicatechin gallate
EGC=(–)-Epigallocatechin
EGCG=(–)-Epigallocatechin gallate
GCG=(+)-Gallocatechin gallate
C=(+)-Catechin
TF=Theaflavins
PBP=Polymeric black tea polyphenol fractions

(XMEs), which broadly can be segregated as Phase I and Phase II enzymes (Yang et al., 1994). Xenobiotic metabolizing enzymes function in the biotransformation (e.g. oxidations, reductions, hydrolysis, hydration, conjugation and condensation) of a wide range of endogenous and exogenous compounds. This metabolism is normally divided into two phases. In the first phase, an electrophilic substituent group is inserted or revealed; in the second phase, the substituent groups are conjugated. Reactive intermediates are produced from non-reactive carcinogen generally by Phase I enzymes (CYP450 monooxygenase system). Due to electrophilicity of the metabolites that are formed, they attack various nucleophilic sites in the tissue and thereby may induce DNA adduct formation, mutation or cell damage. On the other hand, Phase II enzymes

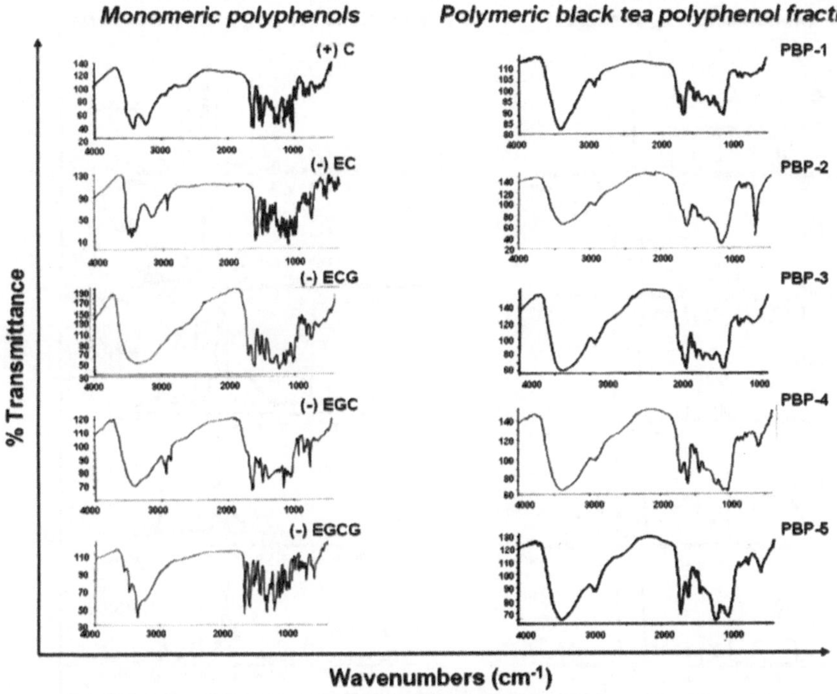

Fig. 9.5 Fourier-transformed infra red (FT-IR) spectra of catechin monomers from green tea and PBPs from black tea

(glutathione S-transferases, sulfatases and UDP glucuronosyl transferases) via trapping the reactive metabolites by glutathione, sulphate or glucuronic acid conjugation, carry out the detoxification of the xenobiotic compounds and their reactive metabolites. Thus the Phase II enzyme induction system is an important component of the cellular stress response in which a diverse array of electrophilic and oxidative toxicants can be removed from the cell before they are able to damage the DNA. Antioxidants and other chemopreventive agents exert their protective effect not only by scavenging reactive oxygen species (ROS), but also by inducing *de novo* expression of genes encoding detoxifying or defensive proteins, including phase II enzymes (Surh, 2003). The 5′ flanking regions of these genes contain a common cis element known as the antioxidant response element (ARE). Many basic leucine zipper (b ZIP) transcription factors, such as NRF, JUN, FOS, FRA, MAF and AH receptor bind to these ARE sequences and modulate the expression of the stress responsive genes (which include some of the Phase I and Phase II enzymes) (Yang et al., 1994; Yu et al., 1997; Lin and Liang, 2000).

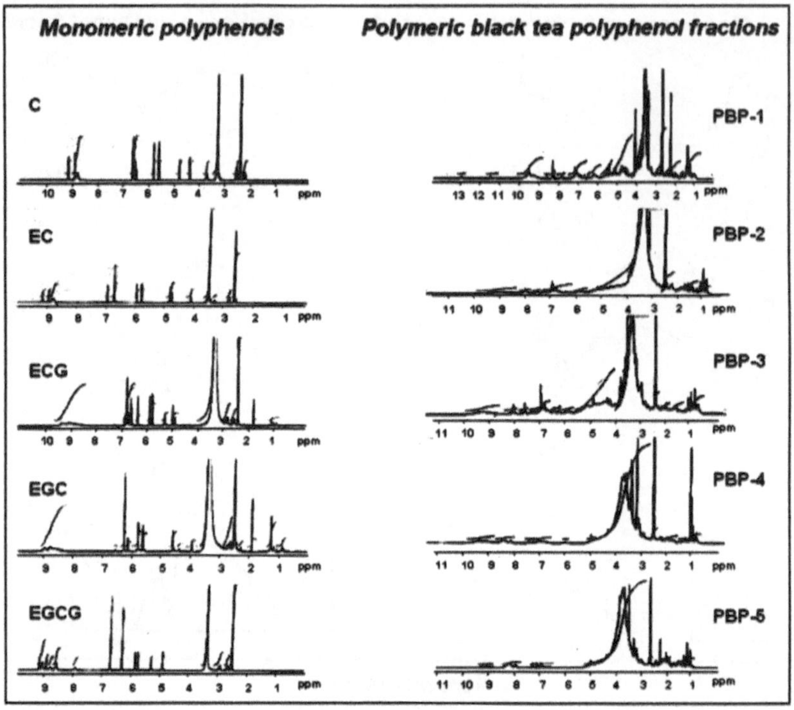

Fig. 9.6 Nuclear Magnetic Resonance (NMR) spectra of catechin monomers from green tea and PBPs from black tea

Table 9.5 Physico-chemical properties of PBPs

Parameters	*PBP-1*	*PBP-2*	*PBP-3*	*PBP-4*	*PBP-5*	*EGCG*	*TFs*
Color	Light brown	Brown	Pale yellow	Dark brown	Dark brown	Orangish brown	Yellowish brown
pH of 1% aq. solutions	5.53	5.55	5.40	4.26	4.09	4.81	5.05
TLC mobility[a]	–	–	–	–	–	+	+
$FeCl_3$ reactivity[b]	Weakly positive	Weakly positive	Weakly positive	Weakly positive	Weakly positive	Strongly positive	Strongly positive
Hydrolysis test[c]	+	+	+	+	+	–	ND
UV λ_{max1}	211	219	217	211	210	215	214
UV λ_{max2}	272	272	268	263	270	273	272
$\lambda_{max1}/\lambda_{max2}$	2.78	1.68	1.78	3.21	6.28	1.45	1.45
FT-IR	Suggestive of polymeric nature					Monomeric	ND
^{1}H-NMR	Suggestive of polymeric nature					Monomeric	ND

a–Sationary phase: Silica gel, Mobile phase: Chloroform: Ethyl acetate: Formic acid (6:4:1)
b–To determine the phenolic nature of test compounds
c–To confirm the proanthocyanidin nature of test compounds

Hence, experiments were conducted to study the effects of decaffeinated black tea extract/PBPs on the activity of Phase I and Phase II enzymes.

Experiments carried out *in vitro* to study the effect of PBPs on the formation of B(a)P-derived DNA adducts show that PBP-1, PBP-2 and PBP-3 (100-800 µg/ml) brought about a dose-dependent decrease in the levels of [^{3}H]-B(a)P-derived DNA adducts. The extent of decrease for PBP-1, PBP-2 and PBP-3 ranged from 13-75%, 7-73% and 19-85% of control values (24.23±0.74 pmol/mg DNA, 100%), respectively. However, inclusion of PBP-4 or PBP-5 (100-800 µg/ml) in the assay did not show significant effect under the experimental conditions used. When rat liver microsomes were pre-incubated with individual PBP-1 to 5 and then incubation mix employed for the assay, a dose-dependent inhibition of B(a)P-derived DNA adducts was observed not only with PBP-1, PBP-2 and PBP-3, but PBP-4 also. The extent of inhibition ranged from 24-100% of control values (24.23±0.74 pmol/mg DNA, 100%). Comparison of extent of inhibition in assays between simultaneous addition (co-incubation with assay mix) of PBPs versus those pre-incubated with microsomes at same dose (closer to IC_{50}) of PBP-1, PBP-2 and PBP-3 showed relatively higher degree or % inhibition of DNA adduct formation with pre-incubation of PBP-1, PBP-2 or PBP-3 with microsomes. Thus PBP-4 showed inhibitory activity only after pre-incubation with microsomes, while PBP-5 failed to show any effect under both the experimental conditions employed (Fig. 9.7).

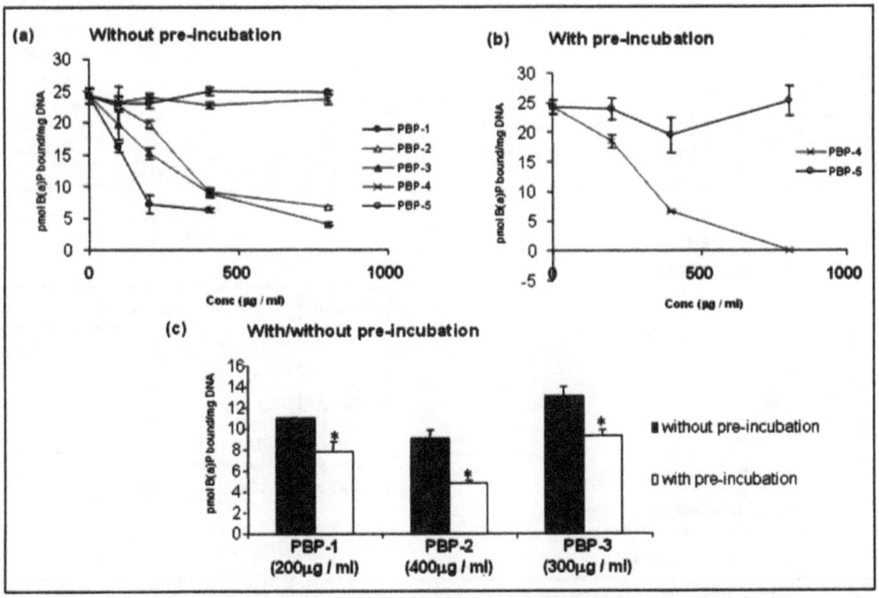

Fig. 9.7 Effect of PBPs on the formation of [^{3}H]-B(a)P-derived DNA adducts *in vitro*

To assess whether inhibition of adduct formation was due to direct scavenging or due to the effect of PBPs on xenobiotic metabolizing enzymes, the activities of CYP450 isozymes were studied. Different isozymes of CYP450, viz., CYP1A1 (EROD), CYP1A2 (MROD) and CYP2B1 (PROD) were studied employing biochemical probes wherein effect of PBPs on the formation of product, i.e., resorufin was measured. The observations illustrate the potent inhibitory effect of PBPs on the activities of specific CYP450 isozymes in Aroclor 1254-treated rat liver microsomes. All PBPs as well as EGCG inhibited the formation of resorufin catalysed by CYP1A1 or 1A2 or 2B1 isozymes from respective substrates in a dose-dependent manner. Relatively lower IC_{50} values were observed for all PBPs and EGCG in inhibiting CYP1A2 (MROD) than CYP1A1 (EROD) and/or CYP2B1 (PROD) activity. Inhibitory effects of all PBPs and EGCG were in the order MROD > EROD > PROD. These results suggest that all PBPs continue to possess CYP450 inhibitory activity as seen with free catechins/EGCG/ green tea polyphenols. Among PBPs, inhibition for MROD was in the order PBP-1 > PBP-3 > PBP-2 > PBP-4 > PBP-5, while inhibition for EROD/PROD followed essentially the same order, i.e., PBP-2 > PBP-1 > PBP-4 > PBP-3 > PBP-5. PBP-5 appeared to be least effective among various PBPs for inhibiting all three isozymes of CYP450 studied (Fig. 9.8).

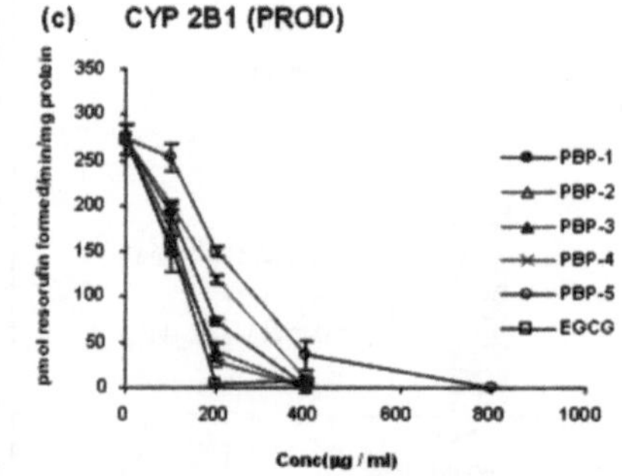

(d)

Fraction	IC_{50} (µg / ml)		
	CYP 1A1	CYP 1A2	CYP 2B1
PBP-1	170	60	160
PBP-2	150	120	150
PBP-3	280	80	340
PBP-4	180	140	200
PBP-5	360	215	710
EGCG	70	50	110

Fig. 9.8 Effect of PBPs on the activity of isozymes of CYP 450 *in vitro*

Critical evaluation of results with PBP-5 suggest that B(a)P appeared to be a relatively more effective inhibitor than PBP-5 in inhibiting the formation of R from ER probably because of relatively higher affinity of microsomes for B(a)P than for PBP-5 (Fig. 9.9). In absence of exact molecular weights of PBP-1 to PBP-5, it is difficult to compare the inhibitory activity at equimolar concentrations.

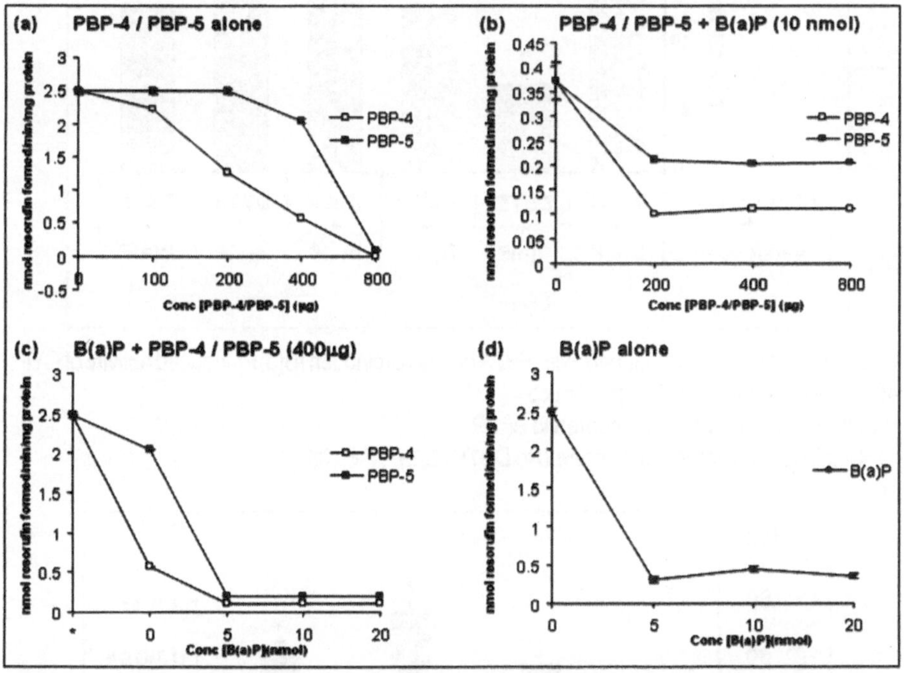

Fig. 9.9 Effect of B(a)P and/or PBPs on the formation of resorufin (R) from ethoxyresorufin (ER)

The above *in vitro* results match very well with *in vivo* studies, which indicate that both 2.5% decaffeinated black tea extract (DBTE) as well as 1% PBP mix (PBPs) inhibited B(a)P-induced activity of CYP1A1 and CYP1A2 in liver as well as lungs of Swiss albino mice (Fig. 9.10).

To study the effect of black tea polyphenols on Phase II enzymes *in vivo*, experiments were carried out where animals were exposed to 1.5% black tea derived TRs/PBPs-rich fraction (BTDTRF) for different time points. Liver and lungs were assayed for total Glutathione *S*-transferase activity (an important Phase II enzyme) by CDNB method. Results indicate that activity of GST was significantly increased in both liver and lungs in tea treated animals. (Fig. 9.11).

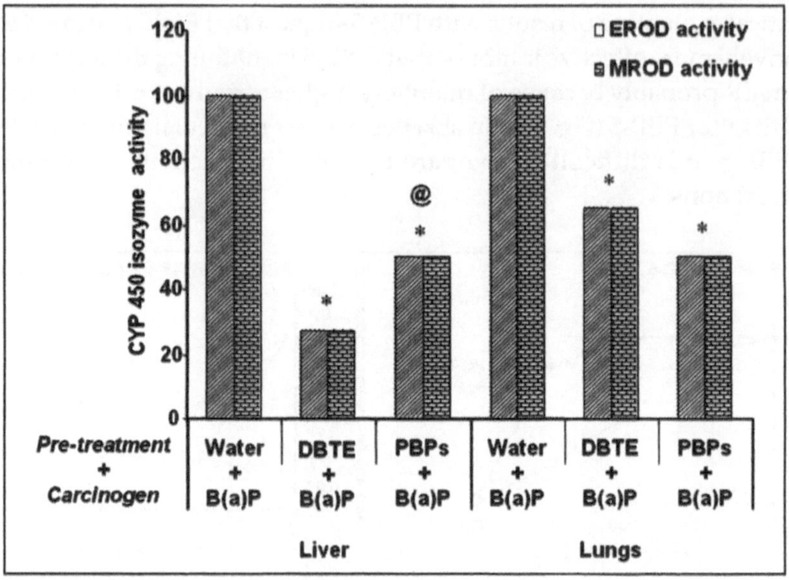

Fig. 9.10 Effect of pre-treatment with PBPs on the carcinogen [B(a)P]-induced activity of CYP 450 isozymes in liver and lungs of mice
*=Significantly different as compared to B(a)P
@=Significantly different as compared to DBTE+B(a)P, $P<0.05$

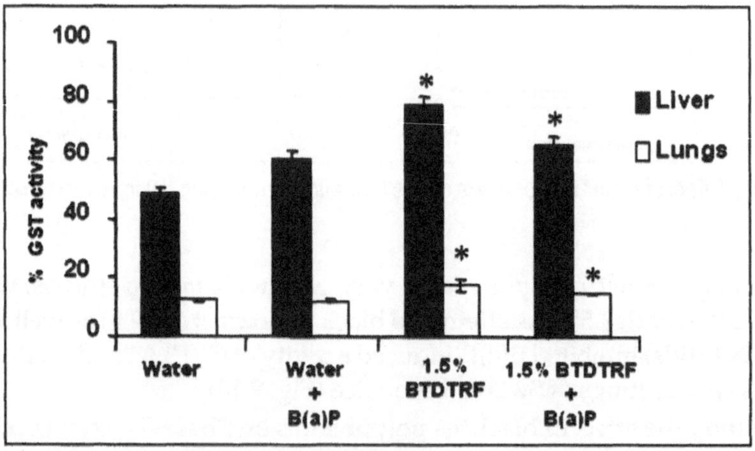

Fig. 9.11 Effect of pre-treatment with 1.5%BTDTRF on GST activity in liver and lungs of mice
*=Significantly different as compared to 'Water', $P < 0.05$

Overall, these observations suggest that black tea polyphenols exhibit anti-initiating activity due to their ability to inhibit CYP450 isozymes (*in vitro* and *in vivo*) and also by inducing the activity of Phase II enzymes like GST (*in vivo*).

The above studies essentially highlight the effectiveness of black tea polyphenol fractions during the early stages of carcinogenesis, i.e., initiation.

Since the process of experimental carcinogenesis is multi-step, i.e., initiation–promotion pathway, investigations were carried out to study the effect of PBP fractions on DMBA-initiated and TPA-promoted skin papillomas in Swiss bare (S/RVCri-ba) mice. Topical application of PBPs mix, prior to that of TPA significantly increased the latency period for the onset of skin papillomas and decreased the multiplicity (average number of skin papillomas per mouse), while the incidence (% animals with skin papillomas) of skin papillomas remained unaffected, after 40 weeks of promotion (Fig. 9.12). Elucidation of mechanism(s) of anti-promoting activity of PBPs revealed down-regulation of ODC activity and decrease in the expression level of gamma glutamyl transpeptidase and proliferating cell nuclear antigen (PCNA) in epidermis of mice treated with PBPs (data not shown).

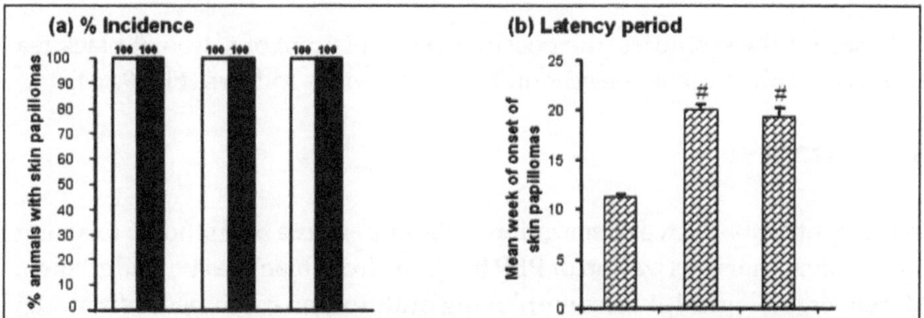

Fig. 9.12 Effect of PBPs mix from black tea and EGCG, the most potent green tea polyphenol on DMBA-initiated and TPA-promoted skin papillomas in Swiss Bare Mice
= Significantly different as compared to TPA
* = Significantly different as compared to TPA at 25 weeks
@ = Significantly different as compared to TPA at 40 weeks of promotion.

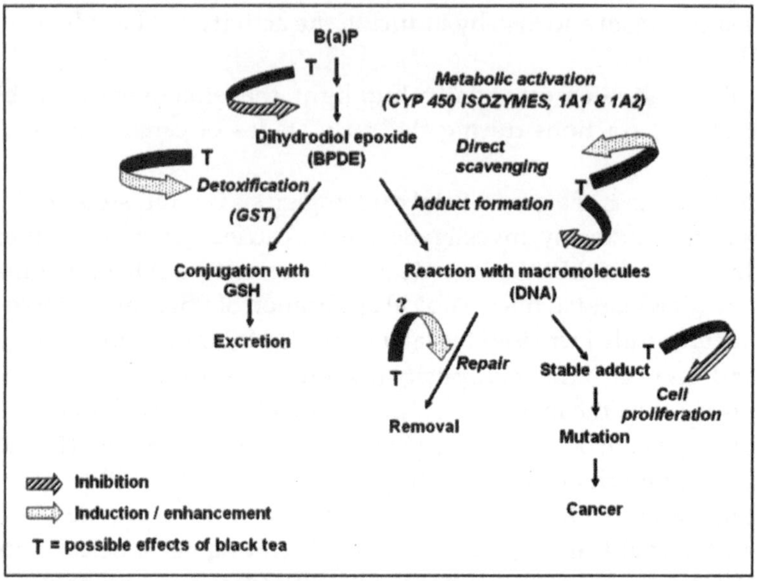

Fig. 9.13 Possible mechanism(s) of black tea-mediated chemoprevention

Based on these studies, the possible mechanism(s) of action of black tea and its polyphenols can be schematically shown as follows (Fig. 9.13).

AN OVERVIEW

In the light of the above observations, it is imperative that studies focusing on chemical characterization of PBP fractions from black tea be undertaken. Moreover, biological studies employing multi-organ carcinogenesis model need to be carried out to ascertain the organ-specific chemopreventive efficacy of black tea. Finally, last but not the least, it becomes important to decipher the molecular mechanism(s) underlying the biological action of black tea and its polyphenols. Chemopreventive efficacy of black tea/ components, if conclusively established in experimental system(s), it could provide a sound background for examining the effects of tea on human health and elucidate appropriate biomarkers for meaningful evaluation of the chemopreventive potential of black tea *in vivo*, in humans.

It must be emphasized here that since the process of carcinogenesis is multi-factorial, any attempt to arrest or reverse the same also needs to be multi-pronged. While keeping the above aspect in mind, avoidance of potentially harmful habits (e.g., cigarette smoking) in concert with a balanced diet, regular exercise, calorie restriction and judicious use of plant-derived antioxidants, can go a long way in either arresting or delaying the onset of cancer (Fig. 9.14). With the advent of mechanism-based chemoprevention and identification of intra-epithelial neoplasia (IEN) as early stage biomarkers for various kinds of cancer, chemoprevention seems to be poised at the threshold of bridging the gap between promise and practicality.

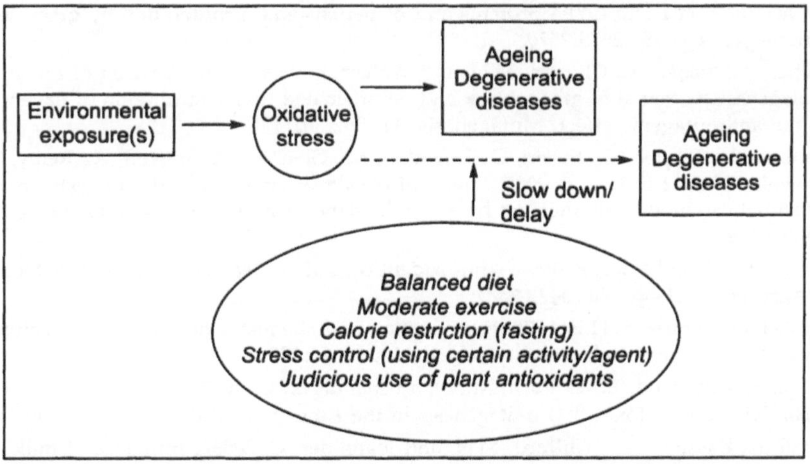

Fig. 9.14 Role of plant-derived antioxidants and lifestyle factors in disease prevention

ACKNOWLEDGEMENTS

The authors thank (a) National Tea Research Foundation (NTRF), Kolkata for partial financial support to the project, (b) Council for Scientific and Industrial Research (CSIR) for the award of Senior Research Fellowship (SRF) to Mr. Rajesh Krishnan (2002-2003), (c) Dr. I.N.N. Namboothiri, Department of Chemistry, IIT, Powai, Mumbai and Dr. L. Mombassawala, Dr. (Mrs.) M. Gandhi and other staff at RSIC, IIT, Powai, Mumbai for FT-IR and NMR analysis/interpretation, (d) Mr. P. Phase, Mr. S. Gosavi and Ms. A. Gawade for excellent technical assistance.

REFERENCES

Agarwal, R., Katiyar, S.K., Khan, S.G. and Mukhtar, H. 1993. Protection against UVB radiation-induced effects in the skin of SKH-1 mice by a polyphenolic fraction isolated from green tea. Photochem. Photobiol., 58, 695-700.

Agarwal, R., Katiyar, S.K., Zaidi, S.I.A. and Mukhtar, H. 1992. SENCAR mice by polyphenolic fraction isolated from green tea and its individual epicatechin derivatives. Cancer Res., 52, 3582-3588.

Bailey, R., Nursten, H. and McDowell, I. 1992. Isolation and analysis of a polymeric thearubigin fraction from tea. J. Sci. Food Agri., 59, 365-375.

Brown, A.G., Eyton, W.B., Holmes, A. and Ollis, W.D. 1969. Identification of the thearubigins as polymeric proanthocyanidins. Nature, 221, 742-744.

Bryce, T., Collier, P.D., Fowlis, I., Thomas, P.E., Frost, D. and Wilkins, C.K. 1970. The structures of theaflavins of black tea. Tetrahed. Lett., 11, 2789-2792.

Bu-Abbas, A., Clifford, M.N., Walker, R. and Ioannides, C. 1994. Selective induction of rat hepatic CYP1 and CYP4 proteins and of peroxisomal proliferation by green tea. Carcinogenesis, 15, 2575-2579.

Bu-Abbas, A., Nanez, X., Clifford, M.N. and Walker, R. 1996. A comparison of the anti-mutagenesis potential of green, black and decaffeinated teas: contribution of flavanols to the antimutagenic effect. Mutagenesis, 11, 597-601.

Caderni, G., De Filippo, C., Luceri, C., Salvadori, M., Giannini, A., Biggeri, A., Remy, S., Cheynier, V. and Dolara, P. 2000. Effects of black tea, green tea and wine extracts on intestinal carcinogenesis induced by azoxymethane in F344 rats. Carcinogenesis, 21, 1965-1969.

Cao, G., Sofic, E. and Prior, R. 1996. Anti-oxidant capacity of tea and common vegetables. J. Agri. Food Chem., 44, 3426-3431.

Cattel, D.J. and Nursten, H.E. 1976. Fractionation and chemistry of ethyl acetate-soluble thearubigins from black tea. Phytochem., 15, 1967-1970.

Catterall, F., Copeland, E., Clifford, M.N. and Ioannides, C. 2003a. Effects of black tea theafulvins on aflatoxin B(1) mutagenesis in the Ames test. Mutagenesis, 18, 145-150.

Catterall, F., Kassimi, A.I., Clifford, M.N. and Ioannides, C. 2003b. Influence of milk on the antimutagenic potential of green and black teas. Anticancer Res., 23, 3863-3867.

Catterall, F., McArdle, N.J., Mitchell, L., Papayanni, A., Clifford, M.N. and Ioannides, C. 2003c. Hepatic and intestinal cytochrome P450 and conjugase activities in rats treated with black tea theafulvins and theaflavins. Food Chem. Toxicol., 41, 1141-1147.

Chan, M., Ho, C.T. and Huang, H. I. 1995. Effects of three dietary phytochemicals from tea, rosemary and turmeric on inflammation-induced nitrite production, Cancer Lett., 96, 23-29.

Chatterjee, M.L., Agarwal, R. and Mukhtar, H. 1996. Ultraviolet B.radiation-induced DNA lesions in mouse epidermis: an assessment using a novel ^{32}P post-labeling technique. Biochem. Biophys. Res. Comm., 229, S90-S95.

Chen, W., Sohn, O.S., Fiala, E.S. and Weisburger, J.H. 1998. Effect of tea on the formation of DNA adducts by azoxymethane. Xenobiotica, 28, 213-217.

Chung, J.Y., Huang, C., Meng, X., Dong, Z. and Yang, C.S. 1999. Inhibition of activator protein 1 activity and cell growth by purified green tea and black tea polyphenols in H-ras-transformed cells: structure–activity relationship and mechanisms involved, Cancer Res., 59, 610-617.

Clark, K.J., Grant, P.G., Sarr, A.B., Belakere, J.R., Swaggerty, C.L., Phillips, T.D. and Woode, G.N. 1998. An *in vitro* study of theaflavins extracted from black tea to neutralize bovine rotavirus and bovine coronavirus infections. Vet. Microbiol., 63, 147-157.

Collier, P.D. and Mallows, R. 1971. Estimation of theaflavins in tea by gas liquid chromatography of their trimethylsilyl ethers. J. Chromat., 57, 19-27.

Constable, A., Varga, N., Richoz, J. and Stadler, R.H. 1996. Antimutagenicity and catechin content of soluble instant teas. Mutagenesis, 11, 189-194.

Coxon, D.T., Holmes, A., Ollis, W.D. and Vora, V.C. 1970. Constitution and configuration of the theaflavin pigments of black tea. Tetrahed. Lett., 11, 5237-5240.

Crispin, D.J., Payne, R.H. and Swaine, D. 1968. Analysis of black tea extracts by chromatography on acetylated Sephadex. J. Chromat., 37, 118-119.

Degenhardt, A., Engelhardt, U.H., Wendt, A-S. and Winterhalter, P. 2000. Isolation of black tea pigments using high-speed countercurrent chromatography and studies on the properties of black tea polymers. J. Agri. Food Chem., 48, 5200-5205.

Degenhardt, A., Engelhardt, U.H., Winterhalter, P. and Ito. Y. 2001. Centrifugal precipitation chromatography – a novel chromatographic system for fractionation of polymeric pigments from black tea and red wine. J. Agri. Food Chem., 49, 1730-1736.

Dhawan, A., Anderson, D., de Pascual-Teresa, S., Santos-Buelga, C., Clifford, M.N. and Ioannides, C. 2002. Evaluation of the antigenotoxic potential of monomeric and dimeric flavanols, and black tea polyphenols against heterocyclic amine-induced DNA damage in human lymphocytes using the Comet assay. Mutat. Res., 515, 39-56.

Dong, Z., Ma, W., Huang, C. and Yang, C.S. 1997. Inhibition of tumor promoter-induced activator protein 1 activation and cell transformation by tea polyphenols, (–)-epigallocatechin gallate, and theaflavins. Cancer Res., 57, 4414-4419.

Ellison, I.F. 2000. Tea and other beverage consumption and prostate cancer risk: a Canadian retrospective cohort study. Eur. J. Cancer Prev., 9, 125-130.

Finger, A., Kuhr, S. and Engelhardt, U.H. 1992. Chromatography of tea constituents. J. Chromat., 624, 293-315.

Fourneau, C., Laurens, A., Hocquemiller, R. and Cave, A. 1996. Radical scavenging evaluation of green tea extracts. Phytother. Res., 10, 529-530.

Frei, B. and Higdon, J.V. 2003. Antioxidant activity of tea polyphenols in vivo: evidence from animal studies. J. Nutr., 133, 3275S-3284S.

Galanis, D.J., Kolonel, L.N., Lee, J. and Nomura, A. 1998. Intakes of selected foods and beverages and the incidence of gastric cancer among the Japanese residents of Hawaii: a prospective study. Int. J. Epidemiol., 27, 173-177.

Gao, Y.T., McLaughlin, J.K., Blot, W.J., Ji, B.T., Dai, Q. and Fraumeni, J.J. 1994. Reduced risk of oesophageal cancer associated with green tea consumption. J. Natl. Cancer Inst., 86, 855-857.

Gardner, P.T., Mcphail, D.B. and Duthie, G.C. 1998. Electron spin resonance spectroscopic assessment of the anti-oxidant potential of teas in aqueous and organic media. J. Sci. Food Agri., 76, 257-262.

Gensler, H.L., Timmermand, B.N., Valcic, S., Wachter, G.A., Dorr, R., Dvorakova, K. and Alberts, D.S. 1996. Prevention of photocarcinogenesis by topical administration of pure epigallocatechin gallate isolated from green tea. Nutr. Cancer, 26, 325-328.

Goldbohm, R.A., Hertog, M.G., Brants, H.A., van Poppel, G. and van den Brandt, P.A. 1996. Consumption of black tea and cancer risk: a prospective cohort study. J. Natl. Cancer Inst., 88, 93-100.

Guo, Q., Zhao, B., Li, M., Shen, S. and Xin, W. 1996. Studies on protective mechanisms of four components of green tea polyphenols against lipid peroxidation in synaptosomes. Biochim. Biophys. Acta, 1304, 210-222.

Gupta, S., Chaudhuri, T., Ganguly, D.K. and Giri, A.K. 2001. Anticlastogenic effects of black tea (World Blend) and its two active polyphenols theaflavins and thearubigins *in vivo* in Swiss albino mice. Life Sci., 69, 2735-2744.

Gupta, S., Chaudhuri, T., Seth, P., Ganguly, D.K. and Giri, A.K. 2002. Antimutagenic effects of black tea (World Blend) and its two active polyphenols theaflavins and thearubigins in Salmonella assays. Phytother. Res., 16, 655-661.

Hammons, G.J., Fletcher, J.V., Stepps, K.R., Smith, E.A., Balentine, D.A., arbowy, M.E. and Kadlubar, F.F. 1999. Effects of chemoprotective agents on the metabolic activation of the carcinogenic arylamines PhIP and 4-aminobiphenyl in human and rat liver microsomes. Nutr. Cancer, 33, 46-52

Harbowy, M. and Balentine, D.A. 1997. Tea chemistry. Crit. Rev. Plant Sci., 16, 415-480.

Hartman, T.J., Tangrea, J.A. and Pietinen, P. 1998. Tea and coffee consumption and risk of colon and rectal cancer in middle-aged Finnish men. Nutr. Cancer, 31, 41-48.

Haslam, E. 2003. Thoughts on Thearubigins. Phytochemistry, 64, 61-73.

Hayakawa, S., Kimura, T., Saeki, K., Koyama, Y., Aoyagi, Y., Noro, T., Nakamura, Y. and Isemura, M. 2001. Apoptosis-inducing activity of high molecular weight fractions of tea extracts. Biosci. Biotechnol. Biochem., 65, 459-462.

Hertog, M.G.L., Feskens, E.J.M., Hollman, P.C.H., Katan, M.B. and Kromhout, D. 1993. Dietary anti-oxidant flavonoids and the risk of coronary heart disease: the Zutphen elderly study. Lancet, 342, 1007-1011.

Hibasami, H., Komiya, T., Achiwa, Y., Ohnishi, K., Kojima, T., Nakanishi, K., Sugimoto, Y., Hasegawa, M., Akatsuka, R. and Hara, Y. 1998. Black tea theaflavins induce programmed cell death in cultured human stomach cancer cells. Int. J. Mol. Med., 1, 725-727.

Hibasami, H., Achinea, Y., Fujikawa, J. and Komiya, T. 1996. Induction of programmed cell death (apoptosis) in human lymphoid leukaemia cells by catechin compounds. Anticancer Res., 16, 1943-1946.

Hirose, M., Hoshiya, T., Akagi, K., Takahashi, S., Hara, Y. and Ito, N. 1993. Effects of green tea catechins in a rat multi-organ carcinogenesis model. Carcinogenesis, 14, 1549-1553.

Hirose, M., Hoshiya, T., Agkagi, K., Futakuchi, M. and Ito, N. 1994. Inhibition of mammary gland carcinogenesis by green tea catechins and other naturally occurring antioxidants in female Sprague-Dawley rats pretreated with 7,12-dimethylbenz(a) anthracene. Cancer Lett., 83, 149-156.

Ho, C.T., Chen, Q., Shi, H., Zhang, K.Q. and Rosen, R.T. 1992. Antioxidative effect of polyphenol extract prepared from various Chinese teas. Prev. Med., 21, 520-525.

Hoefler, A.C. and Coggon, P. (1976) Reverse phase HPLC of tea constituents. J. Chromat., 129, 460-463.

Hong, J., Smith, T.J., Ho, C.T., August, D.A. and Yang, C.S. 2001. Effects of purified green and black tea polyphenols on cyclooxygenase- and lipoxygenase-dependent metabolism of arachidonic acid in human colon mucosa and colon tumor tissues. Biochem. Pharmacol., 62, 1175-1183.

Hong, C.Y., Wang, C.P., Lo, Y.C. and Hsu, F.L. 1994. Effect of flavan-3-ol tannins purified from Camellia sinensis on lipid peroxidation of rat heart mitochondria. Am. J. Chinese Med., 22, 285-292.

Huang, M-T., Ho, C.T., Wang, Z.Y., Ferraro, T., Finnegan-Oliver, T., Lou, Y. R., Mitchell, J.M., Laskin, J.D., Newmark, H., Yang, C.S. and Conney, A.H. 1992. Inhibitory effect of topical application of a green tea polyphenol fraction on tumor initiation and promotion in mouse skin. Carcinogenesis, 13, 947-954.

Huang, M-T., Xie, J.G., Wang, Z.Y., Ho, C.T., Lou, Y. R., Wang, C.X., Hard, G.C. and Conney, A.H. 1997. Effects of tea, decaffeinated tea and caffeine on UVB light-induced complete carcinogenesis in SKH-1 mice: demonstration of caffeine as a biologically important constituent of tea. Cancer Res., 57, 2623-2629.

Imai, K., Suga, K. and Nakachi, K. 1997. Cancer-preventive effects of drinking green tea among a Japanese population. Prev. Med., 26, 769-775.

Inagake, M., Yamane, T., Kitao, Y., Oya, K., Matsumoto, H., Kikuoka, N., Nikatani, H., Takahashi, T., Nishimura, H. and Iwashima, A. 1995. Inhibition of 1,2-

dimethylhydrazine-induced oxidative DNA damage by green tea extract in rat. Jpn. J. Cancer Res., 86, 1106-1111.

Ishikawa, T., Suzukawa, M., Ito, T., Yoshida, H., Ayaori, M., Nishiwaki, M., Yonemura, A., Hara, Y. and Nakamura, H. 1997. Effect of tea flavonoid supplementation on the susceptibility of low-density lipoprotein to oxidative modification. Am. J. Clin. Nutr., 66, 261-266.

Javed, S., Mehrotra, N.K. and Shukla, Y. 1998. Chemopreventive effects of black tea polyphenols in mouse skin model of carcinogenesis. Biomed. Environ. Sci., 11, 307-313.

Ji, B.T., Chow, W.H., Hsing, A.W., McLaughlin, J.K., Dai, Q., Blot, W.J. and Fraumeni, J.F. 1997. Green tea consumption and the risk of pancreatic and colorectal cancers. Int. J. Cancer, 70, 255-258.

Jovanovich, S., Hara, Y., Steenken, S. and Simic, M. 1995. Anti-oxidant potential of gallocatechins. A pulse radiolysis and laser photolysis study. J. Am. Chem. Soc., 117, 9881-9888.

Katiyar, S.K., Agarwal, R., Woods, G.S. and Mukhtar, H. 1992. Inhibition of 12-*O*-tetradecanoylphorbol-13-acetate-caused tumor promotion in 7,12-dimethylbenz(a) anthracene-initiated SENCAR mouse skin by a polyphenolic fraction isolated from green tea. Cancer Res., 52, 6890-6897.

Katiyar, S.K., Agarwal, R., Ekker, S., Wood, G.S. and Mukhtar, H. 1993a. Protection against, TPA caused inflammation in SENCAR mouse ear skin by polyphenolic fraction isolated from green tea. Carcinogenesis, 14, 361-365.

Katiyar, S.K., Agarwal, R. and Mukhtar, H. 1993b. Protective effects of green tea polyphenois administered by oral intubation against chemical carcinogen-induced forestomach and pulmonary neoplasia in A/J mice. Cancer Lett., 73, 167-172.

Katiyar, S.K., Agarwal, R. and Mukhtar, H. 1993c. Protection against malignant conversion of chemically induced benign skin papillomas to squamous cell carcinomas in SENCAR mice by a polyphenolic fraction isolated from green tea. Cancer Res., 53, 5409-5412.

Katiyar, S.K., Agarwal, R. and Mukhtar, H. 1993d. Inhibition of both stage I and stage II skin tumor promotion in SENCAR mice by a polyphenolic fraction isolated from green tea: inhibition depends on the duration of polyphenol treatment. Carcinogenesis, 14, 2641-2643.

Katiyar, S.K., Agarwal, R., Zaim, M.T. and Mukhtar, H. 1993e. Protection against N-nitrosodiethylamine and benzo(a)pyrene-induced forestomach and lung tumorigenesis in A/J mice by green tea. Carcinogenesis, 14, 849-855.

Katiyar, S.K., Elmets, C.A., Agarwal, R. and Mukhtar, H. 1995. Protection against UVB-induced local and systemic suppression of contact hypersensitivity and edema responses in C3H/HeN mice by green tea polyphenols. Photochem. Photobiol., 62, 855-861.

Katiyar, S.K. and Mukhtar, H. 1997a. Tea antioxidants in cancer chemoprevention. J. Cell Biochem. Suppl., 27, 59-67.

Katiyar, S.K. and Mukhtar, H. 1997b. Inhibition of phorbol ester tumor promoter 12-O-tetradecanoylphorbol-13-acetate-caused inflammatory responses in SENCAR mouse skin by black tea polyphenols. Carcinogenesis, 18, 1911-1916.

Kiehne, A. and Engelhardt, U.H. 1996. Thermospray-LC-MS analysis of various groups of polyphenols in tea. II: Chlorogenic acids, theaflavins and thearubigins. Z. Lebensm. Unters. Forsch., 202, 299-302.

Knekt, P., Jarvinen, R., Reunanan, A. and Matela, J. 1996. Flavonoid intake and coronary mortality in Finland: a cohort study. Br. Med. J., 312, 478-481.

Landau, J.M., Wang, Z.Y., Yang, G.Y., Ding, W. and Yang C.S. 1997. Inhibition of spontaneous formation of lung tumors and rhabdomyosarcomas in A/J mice by black and green tea. Carcinogenesis, 19, 501-507.

Lea, M.A., Xiao. Q., Sadhukhan, A.K., Cottle, S., Wang, Z.Y. and Yang, C.S. 1993. Inhibitory effects of tea extracts and (–)-epigallocatechin gallate on DNA synthesis and proliferation of hepatoma and erythroleukaemia cells. Cancer Lett., 68, 231-236.

Lee, M.J., Prabhu, S., Meng, X., Li, C. and Yang, C.S. 2000. An improved method for the determination of green and black tea polyphenols in biomatrices by high-performance liquid chromatography with coulometric array detection. Anal. Biochem., 279, 164-169.

Liang, Y.C., Tsai, D.C., Lin-Shiau, S.Y., Chen, C.F., Ho, C.T. and Lin, J.K. 2002. Inhibition of 12-O-tetradecanoylphorbol-13-acetate-induced inflammatory skin edema and ornithine decarboxylase activity by theaflavin-3,3'-digallate in mouse. Nutr. Cancer, 42, 217-223.

Lin, D.X., Thompson, P.A., Teitel, C., Chen, J.S. and Kadlubar, F.F 2003. Direct reduction of N-acetoxy-PhIP by tea polyphenols: a possible mechanism for chemoprevention against PhIP-DNA adducts formation. Mutat. Res., 523-524, 193-200.

Lin, J.K. and Liang, Y.C. 2000. Cancer chemoprevention by tea polyphenols. Proc. Natl. Sci. Counc. Repub. China, 24, 1-13.

Lodovici, M., Casalini, C., De Filippo, C., Copeland, E., Xu, X., Clifford, M. and Dolara, P. 2000. Inhibition of 1,2-dimethylhydrazine-induced oxidative DNA damage in rat colon mucosa by black tea complex polyphenols. Food Chem. Toxicol., 38, 1085-1088.

Lu, J., Ho, C.T., Ghai, G. and Chen, K.Y 2000. Differential effects of theaflavin monogallates on cell growth, apoptosis, and Cox-2 gene expression in cancerous versus normal cells. Cancer Res., 60, 6465-6471.

Lu, Y.P., Lou, Y.R., Xie, J.G., Yen, P., Huang, M.T. and Conney, A.H. 1997. Inhibitory effect of black tea on the growth of established skin tumors in mice: effects on tumor size, apoptosis, mitosis and bromodeoxyuridine incorporation into DNA. Carcinogenesis, 18, 2163-2169.

Lung, H.L., Ip, W.K., Chen, Z.Y., Mak, N.K. and Leung, K.N. 2004. Comparative study of the growth-inhibitory and apoptosis-inducing activities of black tea theaflavins and green tea catechin on murine myeloid leukemia cells. Int. J. Mol. Med., 13, 465-471.

Luo, M., Wahlqvist, M.L. and O'Brien, R.C. 1997. Inhibition of LDL oxidation by green tea extract. Lancet, 349, 360-361.

Lyn-Cook, B.D., Rogers, T., Yan, Y., Blann, E.B., Kadlubar, F.F. and Hammons, G.J. 1999. Chemopreventive effects of tea extracts and various components on human pancreatic and prostate tumor cells *in vitro*. Nutr. Cancer, 35, 80-86.

Maity, S., Vedasiromoni, J.R. and Ganguly, D.K. 1995. Anti-ulcer effect of the hot water extract of black tea (Camellia sinensis). J. Ethnopharmacol., 46, 167-174.

Maity, S., Vedasiromoni, J.R. and Ganguly, D.K. 1998. Role of glutathione in the antiulcer effect of hot water extract of black tea (*Camellia sinensis*). Jpn. J. Pharmacol., 78, 285-292.

Mangiapane, H., Thomson, J. Salter, A., Brown, S. Bell, D. and White, D.A. 1992. The inhibition of the oxidation of low density lipoprotein by (+)-catechin, a naturally occurring flavonoid. Biochem. Pharmacol., 43, 445-450.

Mazur, W.M., Wahala, K., Rasku, S., Salakka, A., Hase, T. and Adlercreutz, H. 1998. Lignan and isoflavonoid concentrations in tea and coffee. Br. J. Nutr., 79, 37-45.

Michaud, D.S., Spiegelman, D. and Clinton, S.K. 1999. Fluid intake and risk of bladder cancer in men. N. Eng. J. Med., 340, 1390-1397.

Miller, N., Castelluccio, C., Tijburg, L. and Rice-Evans, C. 1996. The anti-oxidant properties of theaflavins and their gallate esters – radical scavengers or metal chelators? FEBS Lett., 392, 40-44.

Misra, A., Chattopadhyay, R., Banerjee, S., Chattopadhyay, D.J. and Chatterjee, I.B. 2003. Black tea prevents cigarette smoke-induced oxidative damage of proteins in guinea pigs. J. Nutr., 133, 2622-2628

Mitsui, T., Yamada, K., Yamashita, K., Matsuo, M., Okuda, A., Kimura, G. and Sugano, M. 1995. E1A-3Y1 cell-specific toxicity of tea polyphenols and their killing mechanism. Int. J. Oncol. 6, 377-383.

Miura, S., Watanabe, J., Sano, M., Tomita, T., Osawa, T., Hara, Y. and Tomita, I. 1995. Effects of various natural antioxidants on the Cu(2+)-mediated oxidative modification of low density lipoprotein. Biol. Pharm. Bull., 18, 1-4.

Miura, S., Watanabe, J., Tomita, T., Sano, M. and Tomita, I. 1994. Inhibitory effects of tea polyphenols (flavan-3-ol derivatives) on Cu^{2+}-mediated oxidative modification of low density lipoprotein. Biol. Pharm. Bull., 17, 1567-1572.

Morel, I., Lescoat, G., Cogrel, P., Sergent, O., Pasdeloup, N., Brissot, P., Cillard, P. and Cilliard, J. 1993. Anti-oxidant and iron-chelating properties of the flavonoids catechin, quercetin and diosmetin on iron-loaded rat hepatocyte cultures. Biochem. Pharmacol., 45, 13-19.

Morse, M.A., Eklind, K.I., Amin, S.G., Hecht, S.S. and Chung, F-L. 1989. Effect of alkyl chain length on the initiation of 4-(methylnitrosamino)-1-(3-pyridyl)-1-butanone-induced lung neoplasia in A/J mice. Carcinogenesis, 10, 1757-1759.

Morse, M.A., Kresty, L.A., Steele, V.E., Kelloff, G.J., Boone, C.W., Balentine, D.A., Harbowy, M.E. and Stoner, G.D. 1997. Effects of theaflavins on N-nitrosomethylbenzylamine-induced esophageal tumorigenesis. Nutr. Cancer, 29, 7-12.

Mukhtar, H., Wang, Z.Y., Katiyar, S.K. and Agarwal, R. 1992. Tea components: Antimutagenic and anticarcinogenic effects. Prev. Med., 21, 351-360.

Nagabhushan, M., Sarode, A.V., Nair, J., Amonkar, A.J., D'Souza, A.V. and Bhide, S.V. 1991. Mutagenicity and carcinogenicity of tea, *Camellia sinensis*. Indian J. Exp. Biol., 29, 401-406.

Nagano, J., Kono, S. and Presto, D.L. 2000. Bladder cancer incidence in relation to vegetable and fruit consumption: a prospective study of atomic bomb survivors. Int. J. Cancer, 86, 132-138.

Nanjo, F., Goto, K., Seto, R., Suzuki, M., Sakai, M. and Hara, Y. 1996. Scavenging effects of tea catechins and their derivatives on 1,1-diphenyl-2-picrylhydrazyl radical. Free Rad. Biol. Med., 21, 895-902.

Narisawa, T. and Fakaura, Y. 1993. A very low dose of green tea polyphenols in drinking water prevents *N*-methyl-*N*-nitrosourea-induced colon carcinogenesis in F344 rats. Jpn. J. Cancer Res., 84, 1007-1009.

Negishi, H., Xu, J.W., Ikeda, K., Njelekela, M., Nara, Y. and Yamori, Y. 2004. Black and green tea polyphenols attenuate blood pressure increases in stroke-prone spontaneously hypertensive rats. J. Nutr., 134, 38-42.

Negre-Salvayre, A., Alomar, Y., Troly, M. and Salvayre, R. 1991. UV treated lipoprotein as a model system for the study of the biological effects of the lipid peroxides in cultured cells, Biochim. Biophys. Acta, 1096, 291-300.

Nishida, H, Omori, M., Fukutomi, Y., Ninomiya, M., Nishiwaki, S., Suganuma, M., Moriwaki, H. and Muto, Y. 1994. Inhibitory effects of (–)-epigallocatechin gallate on spontaneous hepatoma in C3H/HeNCrj mice and human-derived PLC/PRF/5 cells. Jpn. J. Cancer Res., 85, 221-225.

Nomura, M., Ma, W.Y., Huang, C., Yang, C.S., Bowden, G.T., Miyamoto, K. and Dong, Z. 2000. Inhibition of ultraviolet B-induced AP-1 activation by theaflavins from black tea. Mol. Carcinog., 28, 48-55.

Opie, S.C., Robertson, A. and Clifford, M.N. 1990. Black tea thearubigins-their HPLC separation and preparation during in vitro oxidation. J. Sci. Food Agric., 50, 547-561.

Pannala, A., Rice-Evans, C.A., Halliwell, B. and Surinder, S. 1997. Inhibition of peroxynitrite-mediated tyrosine nitration by catechin polyphenols. Biochem. Biophys. Res. Commun., 232, 164-168.

Record, I.R. and Dreosti, I.E. 1998a. Protection by black tea and green tea against UVB and UVA + B induced skin cancer in hairless mice. Mutat. Res., 422,191-199.

Record, I.R. and Dreosti, I.E. 1998b. Protection by tea against UV-A + B-induced skin cancers in hairless mice. Nutr. Cancer, 32, 71-75.

Riemersma, R.A., Rice-Evans, C.A., Tyrrell, R.M., Clifford, M.N. and Lean, M.E. 2001. Tea flavonoids and cardiovascular health. QJM, 94, 277-282.

Roberts, E.A.H., Cartright, R.A. and Oldschool, M., 1957. Fractionation and paper chromatography of water soluble substances from manufactured tea. J. Sci. Food Agric., 8, 72-80.

Roberts, E.A.H. and Smith, R.F. 1961. Spectrophotometric measurements of theaflavins and thearubigins in black tea liquors in assessment of quality of teas. Analyst, 86, 94-98.

Robertson, A. and Bendal, D.S. 1983. Production and HPLC analysis of black tea theaflavins and thearubigins. Phytochem., 22, 883-887.

Saha, P. and Das, S. 2002. Elimination of Deleterious Effects of Free Radicals in Murine Skin Carcinogenesis by Black Tea Infusion, Theaflavins & Epigallocatechin Gallate. Asian Pac. J. Cancer Prev., 3, 225-230.

Salah, N., Miller, N., Paganga, G., Tijburg, L., Bolwell, G.P. and Rice-Evans, C. 1995. Polyphenolic flavanol as scavengers of aqueous phase radicals and as chain-breaking anti-oxidants. Arch. Biochem. Biophys., 322, 339-346.

Sarkar, A. and Bhaduri, A. 2001. Black tea is a powerful chemopreventor of reactive oxygen and nitrogen species: comparison with its individual catechin constituents and green tea. Biochem. Biophys. Res. Commun., 284, 173-178.

Schut, H.A. and Yao, R. 2000. Tea as a potential chemopreventive agent in PhIP carcinogenesis: effects of green tea and black tea on PhIP-DNA adduct formation in female F-344 rats. Nutr. Cancer, 36, 52-58.

Seigler, K. and Ruch, R.J. 1993. Enhancement of gap junctional intercellular communication in tumor promoter treated cells by components of green tea. Cancer Lett., 69, 15-19.

Sengupta, A., Ghosh, S. and Das, S. 2002. Inhibition of cell proliferation and induction of apoptosis during azoxymethane induced colon carcinogenesis by black tea. Asian Pac. J. Cancer Prev., 3, 41-46.

Sengupta, A., Ghosh, S. and Das, S. 2003. Tea can protect against aberrant crypt foci formation during azoxymethane induced rat colon carcinogenesis. J. Exp. Clin. Cancer Res., 22, 185-191.

Shi, S.T., Wang, Z.Y., Smith, T.J., Hong, J.Y., Chen, W.F., Ho, C.T. and Yang, C.S. 1994. Effects of green tea and black tea on 4-(methylnitrosamino)-1-(3-pyridyl)-1-butanone bioactivation, DNA methylation and lung tumorigenesis in A/J mice, Cancer Res., 54, 4641-4647.

Shibata, A., Mack, T.M., Paganini-Hill, A., Ross, R.K. and Henderson, B.E. 1994. A prospective study of pancreatic cancer in the elderly. Int. J. Cancer, 58, 46-49.

Shiraki, M., Hara, Y., Osawa, T., Kumon, H., Nakayama, T. and Kawakishi, S. 1994. Antioxidative and antimutagenic effects of theaflavins from black tea. Mutat. Res., 323, 29-34

Shukla, Y. and Taneja, P. 2002. Anticarcinogenic effect of black tea on pulmonary tumors in Swiss albino mice. Cancer Lett., 176, 137-141.

Siddiqui, I.A., Adhami, V.M., Afaq, F., Ahmad, N. and Mukhtar, H. 2004. Modulation of phosphatidylinositol-3-kinase/protein kinase B- and mitogen-activated protein kinase-pathways by tea polyphenols in human prostate cancer cells. J. Cell. Biochem., 91, 232-242.

Sohn, O.S., Surace, A., Fialaes Richie Jr, J.P., Colosimo, S., Zanag, E. and Weisburger, J.H. 1994. Effects of green and black tea on hepatic xenobiotic metabolizing systems in the male F344 rat. Xenobiotica, 24, 119-127.

Steele, V.E., Kelloff, G.J., Balentine, D., Boone, C.W., Mehta, R., Bagheri, D., Sigman, C.C., Zhu, S. and Sharma, S. 2000. Comparative chemopreventive mechanisms of green tea, black tea and selected polyphenol extracts measured by *in vitro* bioassays. Carcinogenesis, 21, 63-67.

Subramanian, N., Venkatesh, P., Ganguli, S. and Sinkar, V.P. 1999. Role of polyphenol oxidase and peroxidase in the generation of black tea theaflavins. J. Agric. Food Chem., 47, 2571-2578

Surh, Y.J. 2003. Cancer chemoprevention with dietary phytochemicals. Nature Rev. Cancer, 3, 768-780

Taruguchi, S., Fujiki, H., Kobayashi, H., Go, H., Miyado, K., Sadano, H. and Shimokawa, R. 1992. Effect of (–)-epigallocatechin gallate, the main constituent of green tea on lung metastasis with mouse B16 melanoma cell lines. Cancer Lett., 65, 51-54.

Terao, J., Piskula, M. and Yao, Q. 1994. Protective effect of epicatechin, epicatechin gallate and quercetin on lipid peroxidation in phospholipids bilayers. Arch. Biochem. Biophys., 308, 278-284.

Tomita, M., Irwin, K.I., Xie, Z.J. and Santoro, T.J. 2002. Tea pigments inhibit the production of type 1 [T(H1)] and type 2 [T(H2)] helper T cell cytokines in CD4(+) T cells. Phytother. Res., 16, 36-42.

Trevisanato, S.I. and Kim, Y-I. 2000. Tea and Health. Nutr. Rev., 58, 1-10.

Tsubono, Y., Nishino, Y. and Komatsu, S. 2001. Green tea and the risk of gastric cancer in Japan. N. Eng. J. Med., 344, 632-636.

Uchida, S., Ozaki, M., Suzuki, K. and Shikita, M. 1992. Radioprotective effects of (–)-epigallocatechin-3-*O*-gallate (green tea tannin) in mice. Life Sci., 50, 147-152.

Wang, Z.Y., Das, M., Bickers, D.R. and Mukhtar, H. 1988. Interaction of epicatechins derived from green tea with rat hepatic cytochrome P450. Drug Metabol. Dispos., 16, 98-103.

Wang, Z.Y., Khan, W.A., Bickers, D.R. and Mukhtar, H. 1989. Protection against polycyclic aromatic hydrocarbon-induced skin tumor initiation in mice by green tea polyphenols. Carcinogenesis, 10, 411-415.

Wang, Z.Y., Agarwal, R., Bickers, D.R. and Mukhtar, H. 1991. Protection against UV-B radiation-induced photocarcinogenesis in hairless mice by green tea polyphenols. Carcinogenesis, 12, 1527-1530.

Wang, Z.Y., Agarwal, R., Khan, W.A. and Mukhtar, H. 1992a. Protection against benzo(a)pyrene and N-nitrosodiethylamune-induced lung and forestomach tumorigenesis in A/J mice by water extracts of green tea and licorice. Carcinogenesis, 13, 1491-1494.

Wang, Z.Y., Hong, J.Y., Huang, M-T., Reuhl, K.R., Conney, A.H. and Yang, C.S. 1992b. Inhibition of N-Nitrosodiethylamine 1 and 4-(methyinitrosamino)-1-(3-pyridyl)-1-buranone-induced tumorigenesis in A/J mice by green tea and black tea. Cancer Res., 52: 1943-1947.

Wang, Z.Y., Huang, M-T., Ho, C.T., Chang, R., Ma, W., Ferraro, T., Reuhl, K.R., Yang, C.S. and Conney, A.H. 1992c. Inhibitory effect of green tea on the growth of established skin papillomas in mice. Cancer Res. 52; 6657-6665.

Wang, Z.Y., Huang, M-T., Ferraro, T, Wong, C.Q., Lou, Y.R., Reuhl, K., Ioatroupoulos, M., Yang, C.S. and Conney, A.H. 1992d. Inhibitory effect of green tea in drinking water on tumorigenesis by UV light and TPA in the skin of SKH-1 mice. Cancer Res., 52, 1162-1170.

Wang, Z.Y., Huang, M-T., Lou,Y-R., Xei, J.G., Reuhl, K., Newmark, H.L., Ho, C-T., Yang, C.S. and Conney, A.H. 1994. Inhibitory effects of black tea, green tea, decaffeinated black tea and decaffeinated green tea on ultraviolet B light-induced skin carcinogenesis in 7,12-dimethylbenz(a)anthracene-initiated SKH-1 mice. Cancer Res., 54, 3428-3435.

Wang, Z.Y., Wang, L.D., Lee, M.J., Ho, C.T., Huang, M-T. and Yang, C.S. 1995. Inhibition of N-nitrosomethylbenzylamine-induced oesophageal tumorigenesis in rats by green and black tea. Carcinogenesis, 16, 2143-2148.

Wattenberg, L.W. 1987. Inhibitory effects of benzylisothiocyanates administered shortly before diethylnitrosamine or benzo(a)pyrene on pulmonary and stomach neoplasia in A/J mice. Carcinogenesis, 8, 1971-1973.

Wedzicha, B.L. and Donovan, T.J. 1989. Separation of derivatized black tea thearubigins by HPLC. J. Chromat., 478, 217-224.

Wei, H., Zhang, X., Zhao, J., Wang, Z.Y., Bickers, D. and Lebwohl, M. 1999. Scavenging of hydrogen peroxide and inhibition of ultraviolet light-induced oxidative DNA damage by aqueous extracts from green and black teas. Free Radic. Biol. Med., 26, 1427-1435.

Weisburger, J.H. and Chung, F.L. 2002. Mechanisms of chronic disease causation by nutritional factors and tobacco products and their prevention by tea polyphenols. Food Chem Toxicol., 40, 1145-1154.

Weisburger, J.H., Rivenson, A., Garr, K. and Aliaga, C. 1997. Tea, or tea and milk inhibit mammary gland and colon carcinogenesis in rats. Cancer Lett., 114, 323-327.

Wolfram, R.M., Oguogho, A., Efthimiou, Y., Budinsky, A.C. and Sinzinger, H. 2002. Effect of black tea on (iso-)prostaglandins and platelet aggregation in healthy volunteers. Prostaglandins Leukot. Essent. Fatty Acids, 66, 529-533.

Xu, M., Bailey, S.E., Hernaez, J.F., Taoka, C.R., Schut, H.A., and Dashwood, R.H. 1996. Protection by green tea, black tea and indole-3-carbinol against 2-amino-3-methylimidazo [4, 5-f] quinoline-induced DNA adducts and colonic aberrant crypts in F344 rats. Carcinogenesis, 17, 1429-1434.

Xu, Y., Ho, C.T., Amin, S.G., Han, C. and Chung, F.L. 1992. Inhibition of tobacco-specific nitrosamine-induced lung tumorigenesis in A/J mice by green tea and its major polyphenol as antioxidants. Cancer Res., 52, 3875-3879.

Yamane, T., Hagiwara, N., Tateishi, M., Akachi, S, Kim, M., Okuzumi, J., Kitao, Y., Inagake, M., Kuwata, K. and Takahashi, T. 1991. Inhibition of azoxymethane-induced colon carcinogenesis in rat by green tea polyphenol fraction. Jpn. J. Cancer Res., 82, 1336-1339.

Yamane, T., Nakatani, H., Kikuoka, N., Masumoto, H. Aiwata, Y., Kitao, Y., Oya, K. and Takahashi, T. 1996. Inhibitory effects and toxicity of green tea polyphenols for gastro-intestinal carcinogenesis. Cancer, 77, 1662-1667.

Yanagimoto, K., Ochi, H., Lee, K.G. and Shibamoto, T. 2003. Antioxidative activities of volatile extracts from green tea, oolong tea, and black tea. J. Agric. Food Chem., 51, 7396-7401.

Yang, C.S. and Wang, Z.Y. 1993. Tea and cancer. J. Natl. Cancer Inst., 85, 1038-1049.

Yang, C.S., Smith, T.J. and Hong, J.Y. 1994. Cytochrome P450 enzymes as targets for chemoprevention against chemical carcinogenesis and toxicity: opportunities and limitations. Cancer Res., 54 (Suppl.), 1982s-1986s.

Yang, G.Y., Liu, Z., Seril, D.N., Liao, J., Ding, W., Kim, S., Bondoc, F. and Yang, C.S. 1997a. Black tea constituents, theaflavins inhibit 4-(methylnitrosamino)-1-(3-pyridyl)-1-butanone (NNK)-induced lung tumorigenesis in A/J mice. Carcinogenesis, 18, 2361-2365.

Yang, G.Y., Wang, Z.Y., Kim, S., Liao, J., Seril, D., Chen, X, Smth T.J. and Yang, C.S. 1997b. Characterization of early pulmonary hyper-proliferation, tumor progression and their inhibition by black tea in a 4-(methylnitrosamino)-1-(3-pyridyl)-1-butanone (NNK)-induced lung tumorigenesis model with A/J mice. Cancer Res. 57, 1889-1894.

Yang, G.Y., Liao, J., Kim, K., Yurkow, E.J. and Yang, C.S. 1998a. Inhibition of growth and induction of apoptosis in human cancer cell lines by tea polyphenols. Carcinogenesis, 19, 611-616.

Yang, C.S., Yang, G.Y., Landau, J.M., Kim, S. and Liao, J. 1998b. Tea and tea polyphenols inhibit cell hyperproliferation, lung tumorigenesis, and tumor progression. Exp. Lung Res., 24, 629-639.

Yang, G.Y., Liao, J., Li, C., Chung, J., Yurkow, E.J., Ho, C.T. and Yang, C.S. 2000. Effect of black and green tea polyphenols on c-jun phosphorylation and H_2O_2 production in transformed and non-transformed human bronchial cell lines: possible mechanisms of cell growth inhibition and apoptosis induction. Carcinogenesis, 21, 2035-2039.

Yen, G. and Chen, H. 1995. Anti-oxidant activity of various tea extracts in relation to their anti-mutagenicity. J. Agri. Food Chem., 43, 27-32.

Yen, P., Huang, M.-T. and Conney, A.H. 1997. Effect of green or black tea on epidermal carcinogenesis in UVB-treated SKH-1 mice: initiation of tea administration after discontinuation of UVB treatment and prior tumor formation. Proc. Am. Soc. Cancer Res., 38, 367-370.

Yin, P., Zhao, J., Cheng, S., Zhu, Q., Liu, Z. and Zhengguo, L. 1994. Experimental studies of the inhibitory effect of green tea catechin on mice large intestinal cancers induced by 1,2-dimethylhydrazine. Cancer Lett., 79, 33-38.

Yoshino, K., Hara, Y., Sano, M. and Tomita, I. 1994. Antioxidative effects of black tea theaflavins and thearubigin on lipid peroxidation of rat liver homogenates induced by tert-butyl hydroperoxide. Biol. Pharm. Bull., 17, 146-149.

Yu, R., Jias, J.J., Duk, J.L., Gudehithlu, K., Tan, T.H. and Kong, A.N. 1997. Activation of mitogen-activated protein kinases by green tea polyphenols: potential signalling pathways in the regulation of anti-oxidant-responsive element-mediated phase II enzyme gene expression. Carcinogenesis, 18, 451-456.

Zhang, G., Miura, Y. and Yagasaki, K. 2000a. Suppression of adhesion and invasion of hepatoma cells in culture by tea compounds through antioxidative activity. Cancer Lett., 159, 169-173.

Zhang, G., Miura, Y. and Yagasaki, K. 2000b. Induction of apoptosis and cell cycle arrest in cancer cells by in vivo metabolites of teas. Nutr. Cancer, 38, 265-273.

Zhao, J., Jin, X., Yaping, E., Zheng, Z.S., Zhang, Y.J., Athar, M., DeLeo, V.A., Mukhtar, H., Bickers, D.R. and Wang, Z.Y. 1999. Photoprotective effect of black tea extracts against UVB-induced phototoxicity in skin. Photochem. Photobio., 70, 637-644.

Zhao, B., Li, X., He, R., Cheng, S. and Wenjuan, X. 1989. Scavenging effect of extracts of green tea and natural anti-oxidants on active oxygen radicals. Cell Biophys., 14, 75-185.

Chapter

10

Osteoporosis – An Overview: Its Possible Therapeutic Management by Phyto-oestrogens

C. Mitra
Department of Physiology
Presidency College
Kolkata, India

It has been speculated that by the year 2010, the number of patients suffering from osteoporosis will be even greater than it is today, based on current assumptions. The average life span of the elderly segment of population worldwide will increase, and simultaneously the incidence of osteoporosis in this segment will be relatively greater, more so in Southeast Asia and the Pacific Rim (Mundy, 1996).

It is difficult to foresee whether the drugs presently being used for the treatment of osteoporosis will continue to be as efficacious in the next couple of years; or which improved drugs will be used by physicians at that point of time. To review evidence for current therapies for postmenopausal osteoporosis and to establish practical guidelines for management of osteoporosis by family physicians, "medline" was searched from January 1990 to January 2003 (Khan, 2003). According to this report, osteoporosis is treatable. Early diagnosis and intervention is recommended. Approved pharmacological therapies still include alendronate, risedronate, raloxifene, calcitonin and hormone replacement therapy.

INHIBITORS OF BONE RESORPTION

According to their major effects, drugs in this class have been classified on the basis of inhibition of bone resorption and stimulation of bone formation. Drugs in bone resorption class are used primarily for prevention of further bone loss and maintenance of existing bone structure. The most widely used inhibitors of bone resorption are oestrogen, calcitonin and biphosphonates (Mundy, 1996). Continuous efforts have been made to improve the efficacy of current bone resorption inhibitors and to develop new inhibitors of osteoclast action. But, even today when pharmacological therapy is warranted, biphosphonates have shown the greatest benefit in preventing bone loss and lowering fracture rates (Follin and Hansen, 2003).

OESTROGEN

It was first clearly documented about twenty-four years ago that oestrogen is a very effective inhibitor of the rapid phase of bone loss associated with increased bone turnover, which occurs for 5-10 years following menopause. The mechanism by which oestrogens exert their effects on bone resorption has still not been confirmed. It is possible that apart from receptor mediated action, oestrogens exert their effects by modulating the production of local factors in bone like transforming growth factor β (TGF β) and bone resorbing cytokines such as interleukin-6 (IL-6), interleukin-1 (IL-1) and tumour necrosis factor (TNF).

Subsequently, one of the most promising lines of research in the area of bone resorption inhibitors, is in the development of compounds which have oestrogen-like actions on bone mediated through the oestrogen receptor. The goal is a drug with beneficial effects of oestrogen on blood lipids and cardiovascular disease, but without the deleterious effects of oestrogen on the breast and endometrium. A number of compounds with this potential are now under development. The molecular mechanisms by which these interesting compounds may exert agonist effects on the oestrogen receptor in some tissues, but antagonist effects in others, is being clarified. Thus, oestrogen receptor modulators can either have agonist or antagonist effects, depending on the tissue involved. Researchers have classified these oestrogen receptor modulators into four classes, a single class agonists and three distinct classes of antagonists. One of the most important may be raloxifene, which has clearly distinctive properties from droloxifene or tamoxifen.

The success of these compounds as anti-osteoporotic drugs depends on their patient's acceptability, and provided consequently they do not cause irritating side-effects, which are not life-threatening, but nevertheless very uncomfortable for the patient. Non-compliance may also be due to the fact that these agents are being taken as prophylactic drugs for a chronic disease, and patients do not feel any immediate relief or benefit (Mundy, 1996).

CALCITONIN

This probably is equally effective as oestrogen as an inhibitor of osteoclastic bone resorption in the postmenopausal period. However, its widespread use may be fairly limited because of its high cost, effects on inhibition of bone resorption is transient and its effects to increase bone mass are modest (Mundy, 1996).

BIPHOSPHONATES

These compounds are now becoming widely accepted for treatment of increased osteoclastic bone resorption, and particularly for osteoporosis. Though the mechanism of action of biphosphonates is still unclear, it has been suggested that biphosphonates inhibit osteoclast activity by their effects on osteoblasts. Some workers believe that biphosphonates have additional effects on osteoclastic bone resorption which are mediated through immune system (Mundy, 1996).

NEW APPROACHES TO INHIBITION OF BONE RESORPTION

During the last decade, efforts have continued to improve the efficacy of current bone resorption inhibitors and to develop new inhibitors of osteoclast action. Like, for calcitonin, there has been continued improvements in the mode of administration for its better efficacy; for oestrogen, introduction of oestrogen patches to avoid hepatic first-pass metabolism and reduce oestrogenic side-effects; and for biphosphonates, introduction of oral types which are more potent and widely accepted. All the currently available agents (other than calcitonin) have unknown modes of action. Their beneficial effects on inhibition of osteoclastic bone resorption were discovered by serendipity rather than by rational investigation. However, the cellular and molecular mechanisms responsible for bone resorption are now being clarified, and thus further rational therapeutic approaches to inhibiting osteoclast function are possible. As examples, some pharmaceutical companies are investigating inhibitors of tyrosine kinases as drugs for inhibition of bone resorption, because a number of tyrosine kinases are clearly required for both normal osteoclast formation and action. Similarly, a number of adhesion proteins have been linked with osteoclastic bone resorption. These proteins are dependent on the availability of arginine-glycine-aspartic acid (RGD) sequences in attachment proteins. Peptide antagonist for RGD sequences may be effective inhibitors of osteoclastic bone resorption. Likewise, snake venom echistatin has been shown as a powerful inhibitor of osteoclastic bone resorption. Antibodies of E-cadherin molecules also have been suggested as inhibitor of osteoclast formation.

Molecular mechanism by which protons are generated within osteoclasts and then pumped across the ruffled border has also been clarified. These mechanisms involve carbonic anhydrase isoenzyme type-II and a complex vacuolar ATPase which represents the osteoclast proton pump. Inhibitors of both the proton pump mechanism as well as carbonic anhydrase have been identified recently, and these agents inhibit osteoclastic bone resorption *in vitro*.

Finally, proteolytic enzymes which are responsible for bone matrix degradation during osteoclastic bone resorption, including an array of lysosomal enzymes and collagenase, may be inhibited by different agents. These agents include matrix metalloproteinases. Synthetic inhibitors to these proteolytic enzymes may be effective inhibitors of osteoclastic bone resorption.

In the elderly people, there is a decrease in osteoblast function and vigour characterized by a decrease in the capacity of osteoblasts to completely fill in localized resorption defects made by osteoclasts during the remodelling process. This phenomenon is referred to by the bone histomorphometrists as decreased mean wall thickness. This disorder requires an agent which stimulates osteoblastic bone forming activity, probably by increasing their numbers and also their proliferation. Those agents which are currently known to have a stimulatory effect on new bone formation are fluoride, low-dose intermittent PTH and the peptide growth factors which are incorporated into the bone matrix (Mundy, 1996).

OSTEOPOROSIS AND PHYTO-OESTROGENS

In India, we do not have any national health monitoring bureau for prompting and evaluating osteoporosis research. However, according to a recent estimate in the United States, 44 million men and women, in the age group of 50 years or above, have low bone mass or osteoporosis. The most devastating consequence of this disease is bone fractures. The assessment of osteoporosis risk includes determining risk factors, conducting laboratory and physical examinations, and measuring bone mineral density (BMD) and bone-turnover markers. When pharmacological therapy is warranted, biphosphonates and oestrogen-like compounds have shown the greatest benefit in preventing bone loss and lowering fracture rates (Follin and Hansen, 2003).

Nevertheless, there has been an increasing demand for agents for the treatment of postmenopausal symptoms in women with less toxic effects because the currently available agents mentioned in the earlier paragraphs for the treatment of osteoporosis are not fully devoid of adverse effects. Of particular interest in relation to human health, are the class of compounds known as the phyto-oestrogens, which embody several groups of non-steroidal oestrogens including isoflavones and lignans that are widely distributed within the plant kingdom. These phyto-oestrogens are strikingly similar in chemical structure to the mammalian oestrogen, estradiol, and bind to oestrogen receptors (ER) with a preference for the more recently described ER beta (Cassidy, 2003; Duncan et al., 2003). The structural similarity of phyto-oestrogens to endogenous oestrogens has also prompted the hypothesis that phyto-oestrogens exert hormonal or anti-hormonal effects relevant to the risk of hormone-dependent disease and/or their suitability as a dietary alternative to hormone replacement therapy (Duncan et al., 2003). Numerous other biological effects independent of the ER (e.g. antioxidant capacity, antiproliferative and antiangiogenic effects) have been ascribed to these compounds (Cassidy, 2003). The limited studies conducted so far in humans clearly confirm that soya isoflavones can exert hormonal effects (Cassidy, 2003). These effects may be of benefit in the prevention of many of the common diseases observed in Western populations (such as breast cancer, prostate cancer, menopausal symptoms, osteoporosis) where the diet is typically devoid of these biologically active naturally occurring compounds (Cassidy, 2003). A recent study has shown that Epimedium total flavonoids phytosomes (EFP) can improve the bone density, enhance estrogen 2 (E2) level and decrease the IL-6 concentration in serum of ovariectomized rats (Jiang et al., 2002). Laboratory research and epidemiological findings have also revealed that natural isoflavons and polyphenols play a protective role against the development of numerous chronic diseases, including several cancers, cardiovascular disease and osteoporosis (Cassidy, 2003; Duncan et al., 2003; Jiang et al., 2002; Arjmandi et al., 1996; Weishburger, 1996; Toshiya et al., 1999; Hegarty et al., 2000).

Black tea, one of the most favored beverages all over the world, has been well documented to have health-promoting action because of its well-defined polyphenols, theaflavine gallate, digallate and thearubigens (Weisburger, 1996). However, research underlying the beneficial health effects of flavonoids and polyphenols, particularly with tea and osteoporosis, is fairly recent. The American Journal of Clinical Nutrition (April, 2000) published an epidemiological study which suggests that nutrients found in tea (e.g., flavonoids) may positively influence bone mineral density (BMD) and that tea drinking may protect the body's defenses against osteoporosis in older women (Hegarty, 2000).

Being influenced by this notion of black tea action, a study was initiated by a group of Indian researchers to evaluate the phyto-oestrogenic potentiality of black tea extract (BTE) in an oophorectomized rat model (Das et al., unpublished data).

In this study, the oestrogenic potential of whole aqueous extract of Black tea (BTE) (2.5%, 1 ml/100 gm body weight/day for 21 days) was examined on bilaterally oophorectomized rats. BTE supplementation for 21 days in bilaterally oophorectomized rats revived normal reproductive cyclicity. Also, vaginal smear showed re-appearance of cellular features from diestrous only to normal, i.e., all the oestrous stages re-appeared with normal sequence of occurrence (Fig. 10.1).

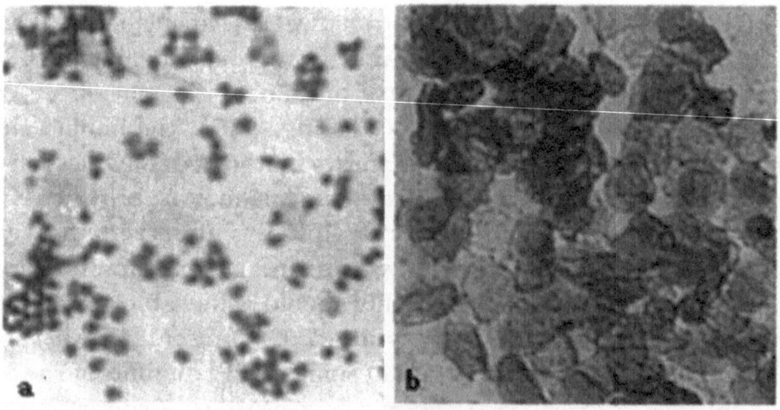

(a) OVX-induced osteoporosis : Rat uterine cell features (Diestrous)

(b) Recovery of OVX-induced osteoporosis by BTE: Rat uterine cell features (Estrous)

Fig. 10.1 Effect of BTE on restoration of normal estrous stage in oophorectomized rat. In oophorectomized rat, the feature was predominatly diestrous (a) On BTE supplementation (2.5%, 1 ml/100 gm body weight/day for 21 days) estrous (b) revived.

BTE supplementation also could restore serum oestrogen titre of these animals. To explore whether such restoration of oestrogen level has any preventive role upon the oestrogen deficiency-related decay of skeletal health, marker parameters of bone resorption and osteoclastic activity (tartrate-resistant acid phosphatase), collagen degradation (urinary hydroxyproline), bone loss (bone ash mineral content) and bone breaking strength (bone density) were examined.

Results indicated that BTE supplementation could significantly reduce oophorectomy-induced skeletal health decaying changes (Table 10.1), suggesting its potentiality as an anti-osteoporotic agent. Since all these results of this study were directly or indirectly linked with preservation and/or maintenance of skeletal health, most likely BTE and skeletal health are associated. Furthermore, Spearman's rank correlation coefficient analysis in between serum oestrogen, bone density and bone ash content of minerals (Table 10.2) support this view.

In another study with oophorectomized rats (ovx), the same group of researchers re-examined the phyto-oestrogenic potentiality of BTE (Das et al., 2004). Results of body weight study revealed that rats in the oophorectomized group had significantly higher final body weight than rats in the sham group. This higher final body weight was not observed in animals receiving BTE. The oophorectomized group also had significantly higher abdominal fat mass and liver weight; and significantly lower uterus, lower right kidney and left kidney weights than in other two groups. All these organ weight changes in oophorectomized group also were not observed in animals receiving BTE (Table 10.3).

Results of urinary studies revealed that rats in the oophorectomized group had significantly higher urinary excretion of calcium (Ca), phosphate, creatinine (Cr), calcium to creatinine (Ca : Cr) ratio ($P<0.001$) and hydroxyproline (HPr) ($P<0.01$) than rats in the sham group. Significant recovery of all these parameters were observed in animals receiving BTE (Table 10.4).

The oophorectomized group also had significantly higher ($P<0.001$) serum alkaline phosphatase (AP) (Fig. 10.2) and tartrate-resistant acid phosphatase (TRAP) (Fig. 10.3) activity than rats in the other two groups. These changes also could not be seen in animals receiving BTE.

Also, identical changes were seen in bone density experiments. Rats in the oophorectomized group had significantly lower densities of the right femur ($P<0.001$), eighth thoracic rib ($P<0.001$), eighth thoracic vertebra ($P<0.05$), and fourth lumbar vertebra ($P<0.01$) than rats in the sham group; and significant improvement in densities of these bones were seen in animals supplemented with BTE.

Table 10.1 Effect of BTE on oophorectomy-induced changes in serum estrogen, tartrate-resistant acid phosphatase, urinary hydroxyproline, bone densities, bone calcium and bone phosphate levels. (Control : Group A, oophorectomized : Group B, and oophorectomized supplemented with BTE : Group C, n = 7 for each group)

				*Significance Level	**Significance Level		
	Control	*Oophorectomized*	*Oophorectomy + BTE*		*Group A vs Group B*	*Group B vs Group C*	*Group A vs Group C*
Serum estrogen (pg/ml)	73.69 ± 2.37	14.50 ± 1.46	38.24 ± 6.66	P<0.001	P<0.01	P<0.01	P<0.01
Serum TRAP (U/L)	0.858 ± 0.067	1.406 ± 0.098	0.866 ± 0.053	P<0.01	P<0.01	P<0.01	NS
Urinary hydroxyproline (mg /24 h)	0.342 ± 0.035	0.725 ± 0.064	0.378 ± 0.059	P<0.01	P<0.01	P<0.01	NS
Bone densities (gm/cm^3)							
Femur	1.430 ± 0.384	1.191 ± 0.022	1.314 ± 0.019	P<0.001	P<0.01	P<0.01	P<0.01
Thoracic rib	1.628 ± 0.050	1.330 ± 0.023	1.517 ± 0.028	P<0.001	P<0.01	P<0.01	NS
Thoracic vertebra	1.371 ± 0.035	1.188 ± 0.030	1.279 ± 0.007	P<0.001	P<0.01	P<0.01	NS
Lumbar vertebra	1.251 ± 0.010	1.145 ± 0.016	1.254 ± 0.023	P<0.01	P<0.01	P<0.01	NS
Bone calcium (% of ash weight)							
Femur	23.37 ± 0.49	18.92 ± 0.51	23.00 ± 0.69	P<0.01	P<0.01	P<0.01	NS
Thoracic rib	36.45 ± 0.70	24.90 ± 0.25	32.04 ± 1.18	P<0.001	P<0.01	P<0.01	P<0.01
Thoracic vertebra	22.01 ± 0.42	11.84 ± 0.55	21.31 ± 0.15	P<0.001	P<0.01	P<0.01	NS
Lumbar vertebra	21.09 ± 0.77	17.64 ± 0.40	21.38 ± 0.67	P<0.01	P<0.01	P<0.01	NS
Bone phosphate (% of ash weight)							
Femur	20.24 ± 0.46	18.11 ± 0.52	21.57 ± 0.28	P<0.001	P<0.01	P<0.01	NS
Thoracic rib	23.33 ± 0.21	20.80 ± 0.43	22.64 ± 0.24	P<0.001	P<0.01	P<0.01	NS
Thoracic vertebra	20.54 ± 0.22	17.16 ± 0.52	20.98 ± 0.08	P<0.001	P<0.01	P<0.01	NS
Lumbar vertebra	20.50 ± 0.45	16.50 ± 0.20	20.29 ± 0.47	P<0.01	P<0.01	P<0.01	NS

*Significance based on Kruskal-Wallis test.

**Significance based on Mann-Whitney 'U' multiple comparison test.

NS denotes not significant.

Table 10.2 Correlations of serum estrogen level with bone ash mineral contents (calcium and phosphate) and bone densities of control group (Group A), oophorectomized group (Group B), and oophorectomized group supplemented with BTE (Group C) of rats (n=7 for each group).

	Control		*Oophorectomized*		*Oophorectomy + BTE*	
	*r value***	*Significance Level**	*r value***	*Significance Level**	*r value***	*Significance Level**
Bone calcium (% of ash weight)						
Femur	+0.866	P<0.05	+0.929	P<0.01	+0.929	P<0.01
Thoracic rib	+0.902	P<0.01	+0.929	P<0.01	+0.929	P<0.01
Thoracic vertebra	+0.902	P<0.01	+0.884	P<0.01	+0.839	P<0.05
Lumbar vertebra	+0.911	P<0.01	+0.938	P<0.01	+0.929	P<0.01
Bone phosphate (% of ash weight)						
Femur	+0.902	P<0.01	+0.929	P<0.01	+0.911	P<0.01
Thoracic rib	+0.955	P<0.001	+0.911	P<0.01	+0.893	P<0.01
Thoracic vertebra	+0.902	P<0.01	+0.929	P<0.01	+0.929	P<0.01
Lumbar vertebra	+0.920	P<0.01	+0.929	P<0.01	+0.893	P<0.01
Bone densities (gm/cm^3)						
Femur	+0.875	P<0.05	+0.893	P<0.01	+0.938	P<0.01
Thoracic rib	+0.955	P<0.001	+0.893	P<0.01	+0.875	P<0.05
Thoracic vertebra	+0.902	P<0.01	+0.920	P<0.01	+0.929	P<0.01
Lumbar vertebra	+0.902	P<0.01	+0.920	P<0.01	+0.911	P<0.01

*Significance based on Spearman's rank correlation coefficient test.

**denotes Spearman's rank difference correlation coefficient.

Table 10.3 Body weight and relative organ weights of sham group (Group A), oophorectomized (ovx) group (Group B), oophorectomized + BTE treated group (Group C) of rats.

Parameters	*Experimental groups*			*Significance level*	
	Sham (Gr. A)	*Ovx (Gr. B)*	*Ovx +BTE (Gr.C)*	*Gr. A vs Gr. B*	*Gr. B vs Gr. C*
Initial body weight (g)	94.4 ± 1.78	97.4 ± 3.4	96.08 ± 1.8	P>0.05	P> 0.05
Final body weight (g)	96.3 ± 1.82	117.8 ± 4.21	97.9 ± 1.23	P<0.01	P< 0.01
Organ weight (g/100 g body weight)					
Uterus	0.1572 ± 0.003	0.0767 ± 0.006	0.1482 ± 0.008	P<0.001	P<0.001
Abdominal fat	1.1624 ± 0.074	2.2097 ± 0.296	1.1051 ± 0.112	P<0.01	P<0.01
Liver	3.27 ± 0.116	4.14 ± 0.282	3.35 ± 0.196	P<0.05	P<0.05
Right kidney	0.3474 ± 0.004	0.3149 ± 0.006	0.3403 ± 0.007	P<0.01	P<0.05
Left kidney	0.3436 ± 0.005	0.2926 ± 0.005	0.3479 ± 0.02	P<0.001	P<0.01

Values are expressed as mean ± SE (n = 5).

Table 10.4 Urinary excretion of calcium, phosphate, creatinine, hydroxyproline and calcium: creatinine ratio in sham group (Group A), oophorectomized (ovx) group (Group B) and oophorectomized + BTE treated group (Group C) of rats.

Parameters	*Experimental groups*			*Significance level*	
	Sham (Gr. A)	*Ovx (Gr. B)*	*Ovx + BTE (Gr. C)*	*Gr. A vs Gr. B*	*Gr. B vs Gr. C*
Calcium (mg/24 h)	0.61 ± 0.029	2.36 ± 0.12	0.81 ± 0.029	P<0.001	P<0.001
Phosphate (mg/24 h)	2.26 ± 0.093	3.32 ± 0.093	2.75 ± 0.070	P<0.001	P<0.01
Creatinine (mg/24 h)	0.79 ± 0.017	1.57 ± 0.12	0.84 ± 0.018	P<0.001	P<0.001
Hydroxyproline (mg/24 h)	0.312 ± 0.05	0.716 ± 0.08	0.332 ± 0.08	P<0.01	P<0.01
Calcium : Creatinine	0.603 ± 0.025	1.644 ± 0.13	1.06 ± 0.089	P<0.001	P<0.01

Values are expressed as mean ± SE (n = 5).

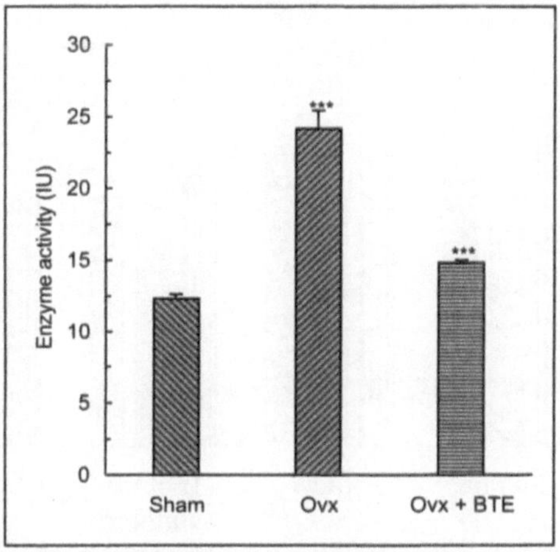

Fig. 10.2 Effect of oophorectomy (ovx) and oophorectomy + BTE on serum alkaline phosphatase (AP) activity in different groups of rats. Error bars represent means ± SE (n=6). In statistical analysis, Group B (ovx) has been compared with Group A (sham-control) and Group C (ovx + BTE) with Group B (ovx). Significance based on Students *t*- test.
*** denotes significant difference P<0.001.

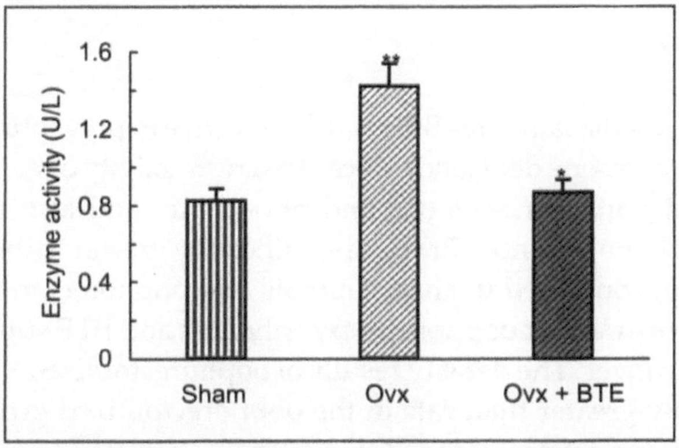

Fig. 10.3 Effect of oophorectomy (ovx) and oophorectomy + BTE on serum tartrate-resistant acid phosphatase (TRAP) activity in different groups of rats. Error bars represent means ± SE (n=6). In statistical analysis Group B (ovx) has been compared with Group A (sham-control) and Group C (ovx + BTE) with Group B (ovx). Significance based on Student's *t*- test.
* denotes significant difference P<0.01
** denotes P<0.001.

Animals of oophorectomized group also showed significant decrease in calcium and phosphate level in all these bones which could be regained significantly when these animals were supplemented with BTE (Table 10.5).

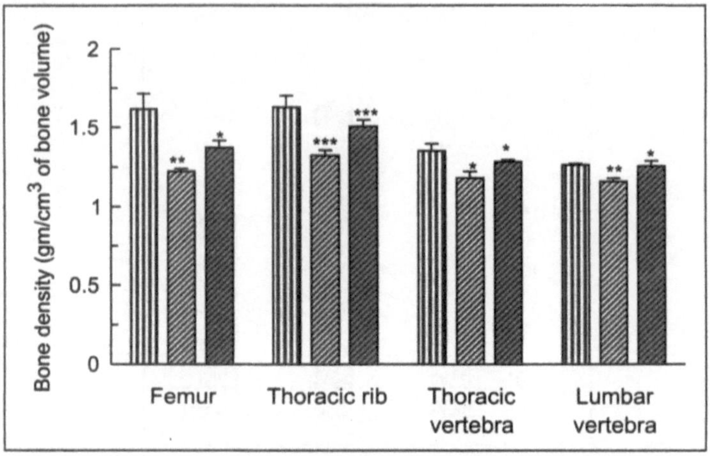

Fig. 10.4 Effect of oophorectomy and oophorectomy + BTE on densities of right femur, 8th thoracic rib, 8th thoracic vertebra, and 4th lumbar vertebra in (▥) Group A (sham-control), (▨) Group B (ovx) and (■) Group C (ovx + BTE) of rats. Error bars represent means ± SE (n=6). In statistical analysis Group B has been compared with Group A and Group C with Group B. Significance based on Student's *t*-test.
* denotes significant difference $P<0.05$,
** denotes $P<0.01$,
*** denotes $P<0.001$.

Results suggest that aqueous BTE may be effective in preventing bone loss due to ovarian hormone deficiency. Because serum activity of AP, TRAP and urinary loss of bone minerals (Ca and phosphate) and also the organic components of bone (Cr and HPr) were significantly greater in the oophorectomized group, compared to sham animals and oophorectomized + BTE group. This confirms that oophorectomy enhances and BTE suppresses the rate of bone turnover. The density results of oophorectomized + BTE group are significantly greater than rats in the oophorectomized group, further suggesting that formation exceeded resorption.

CONCLUDING REMARKS

There has been an unabated search for an alternative of currently available inhibitors of osteoporosis with less side-effects. BTE, from that point of view, may be therapeutically assessed as a phytoestrogenic compound for prevention against oophorectomy-induced osteoporotic damages. The development of BTE or its individual polyphenols as inhibitors of hypogonadal osteoporosis will need further detailed studies, as also its dose and human

Table 10.5 Bone calcium and bone phosphate levels of the sham group (Group A), oophorectomized (ovx) (Group B), and oophorectomized + BTE treated group (Group C) of rats.

	Experimental groups			*Significance level*	
	Sham (Gr. A)	*Ovx (Gr. B)*	*Ovx +BTE (Gr. C)*	*Gr. A vs Gr. B*	*Gr. B vs Gr. C*
Bone calcium level (percent of ash weight)					
Right femur	23.16 ± 0.64	18.61 ± 0.73	22.94 ± 0.83	P<0.001	P<0.01
Eighth thoracic rib	35.49 ± 1.57	24.56 ± 0.31	32.87 ± 1.85	P<0.001	P<0.01
Eighth thoracic vertebra	21.48 ± 0.80	11.30 ± 0.91	21.08 ± 0.34	P<0.001	P<0.001
Fourth lumbar vertebra	21.00 ± 1.24	17.15 ± 0.40	20.23 ± 1.03	P<0.05	P<0.05
Bone phosphate level (percent of ash weight)					
Right femur	20.08 ± 0.58	17.72 ± 0.84	21.15 ± 0.69	P<0.05	P<0.01
Eighth thoracic rib	22.85 ± 0.37	20.78 ± 0.51	22.31 ± 0.34	P<0.01	P<0.05
Eighth thoracic vertebra	20.31 ± 0.40	17.53 ± 0.82	20.73 ± 0.32	P<0.05	P<0.01
Fourth lumbar vertebra	20.29 ± 0.59	16.84 ± 0.47	20.95 ± 0.18	P<0.01	P<0.001

Values are expressed as mean ± SE (n=6).

efficacy extrapolation. Nevertheless, it is an exciting challenge full of promising prospects without any possible side-effects.

REFERENCES

Arjmandi, B.H., L. Alekel, B.W. Hollis, D. Amin, M. Stacewicz-sapuntzakis, P. Guo and S.C. Kukreja. 1996. Dietary soybean protein prevents bone loss in an ovariectomized rat model of osteoporosis. J. Nutr. 126: 161-167.

Cassidy, A. 2003. Potential risks and benefits of phytoestrogen-rich diets. Int. J. Vitam. Nutr. Res. 73(2): 120-126.

Das, A.S., M. Mukherjee, and C. Mitra. 2004. Evidence for a prospective anti-osteoporosis effect of Black Tea (*Camellia sinensis*) extract (BTE) in a bilaterally ovariectomized rat model. Asia Pacific J. Clin. Nutr. 13(2): 210-216.

Das, A.S., D. Das, M. Mukherjee, S. Mukherjee and C. Mitra. Phytoestrogenic effects of black tea (*Camellia sinensis*) extract in an oophorectomized rat (*Rattus norvegicus*) model of osteoporosis. (Unpublished data.)

Duncan, A.M., W.R. Phipps and M.S. Kurzer. 2003. Phyto-oestrogens. Best. Pract. Res. Clin. Endocrinol. Metab. 17(2): 253-71.

Follin, S.L. and L.B. Hansen. 2003. Current approaches to the prevention and treatment of postmenopausal osteoporosis. Am. J. Health Syst. Pharm. 60(9): 883-901.

Hegarty, V.M., H.M. May and K.T. Khaw. 2000. Tea drinking and bone mineral density in older women. Am. J. Clin. Nutr. 71(4): 1003-1007.

Jiang, Y.N., H.Y. Mo and J.M. Chen. 2002. Effects of epimedium total flavonoids phytosomes on preventing and treating bone-loss of ovariectomized rats. Zhongguo. Zhong. Yao. Za. Zhi. 27(3): 221-4.

Khan, A. 2003. Advances in osteoporosis therapy. 2003 update of practical guidelines. Can. Fam. Physician 49: 441-447.

Mundy, G.R. 1996. Osteoporosis into the year 2010. Br. J. Obstet. Gynaecol.. 103(13): 32-38.

Toshiya, T., U. Takehiko, H. Kuniaki, N. Haruo, T. Kuniro and I. Hitoshi. 1999. New 6-0-Acylisoflavone glycosides from soybeans fermented with *Bacillus subtilis (natto)*. I. 6-0-succinylated isoflavone glycosides and their preventive effects on bone loss in ovariectomized rats fed a calcium-deficient diet. Biol. Pharma. Bull. 22(11): 1193-1201.

Weisburger, J.H. 1996. Tea antioxidants and health. In: Handbook of Antioxidants. Cadenas E. and Packer L. (eds). Marcel-Dekker, Inc. N.Y. pp. 469-486.

Chapter 11

Tea and Health: Beyond Hype

B. Banerjee
National Tea Research Foundation
Kolkata, India

Tea currently is the hot topic in both nutritional and therapeutic research worldwide. This is not so because tea is the most preferred drink after water (Banerjee, 1988), but because of the presence of crucial therapeutic compounds in tea which are more bio-stable and direct acting than those found in other plants. The activities of these compounds are so all pervading that they are virtually broad spectrum in their actions (Weisburger, 2000). Besides, the natural integration of aromatic and therapeutic compounds in tea is a rather unique attribute (Banerjee, 1988).

THE CORE COMPOUNDS

Flavonoids generate powerful antioxidant activities which are basically responsible for the therapeutic effects of both of black and green teas, as well as for their deliciously rich flavour and colour. As light is essential for the synthesis of flavonoids, they are common only on those parts of plants and fruits which are optimally exposed to sunlight. They are common on the outer skin of fruits, outer leaves of cabbage and on the leaves of tea. While the outer peels of fruits and vegetables are generally discarded before consumption, it is not so in the case of tea. Hence, the singular important aspect of tea as an exceptional and rich source of polyphenols (like flavonoids) which possess health promoting potential, makes it a beneficial drink. Flavonoids have rich chemical diversity and the level of their presence in tea can be classified as rich, moderate or lower on the following scale (Table 11.1).

Table 11.1 Quantitative variation of flavonoid content in tea

Grade	*Quantity*
Rich	>10 mg/100 gm (Flavanols)
Moderate	4.0 - 9.9 mg/100 gm
Low	< 3.9 mg/100 gm (Flavonols)

For tea, flavonoid is a generic term and a hierarchical arrangement of the compounds derived from it (Table 11.2) shows the overall interrelationships between them.

Table 11.2 Classification of tea polyphenols

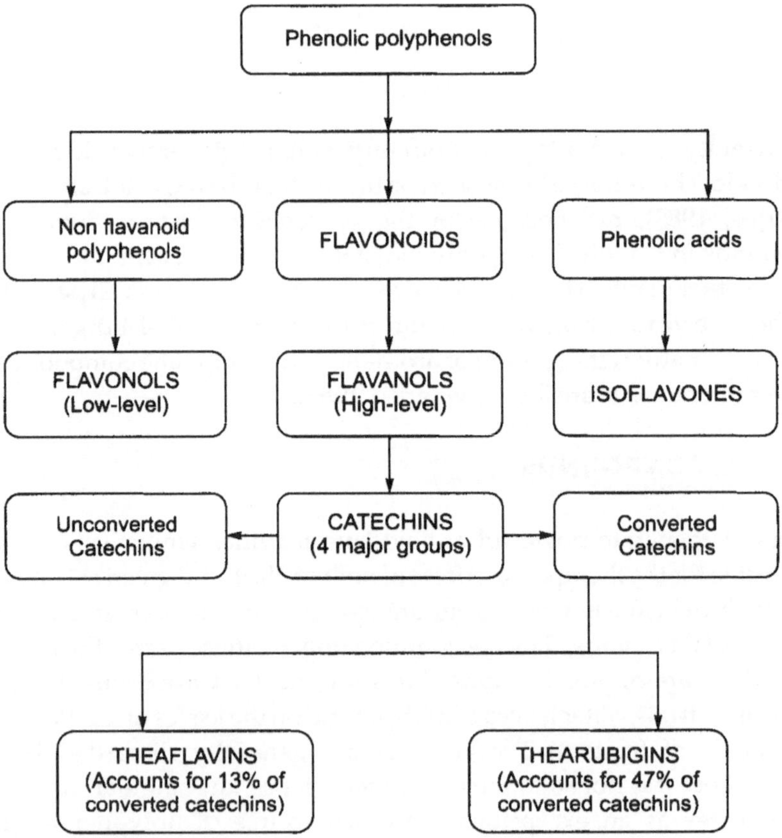

N.B. Compounds in CAPITALS are of medicinal importance. Also refer to Appendix I.

FLAVONOIDS IN TEA INFUSION

About 30% of the solids extracted in a typical black tea infusion are the flavonoids, of which 10% are water soluble flavonols. Most of the remaining flavonoids are the flavanols present as catechins or their derivatives. During the processing of black tea, these catechins are oxidized to produce theaflavins or thearubigins, depending on the colour they take, but they maintain the basic structural configuration ($C_6 - C_3 - C_6$) of the flavanoids. The distribution of generic flavonoids and their specific products in green and black tea however varies (Table 11.3).

Table 11.3 Distribution (per cent) of flavonoids in green and black tea

Compounds	*Green tea*	*Black tea*
Flavonols	10%	10%
Flavanols		
Unconverted catechins	90%	30%
Converted catechins		
(a) Thearubigins		
(b) Theaflavins	—	47%
	—	13%
	100%	100%

Interestingly enough, despite this compositional difference, green and black teas are alike in their therapeutic effects, as emphasized by many papers in this volume. However, the flavanoid level of tea depends on a number of physical and biological factors and their interactions (Table 11.4).

Table 11.4 Factors influencing flavonoid composition of tea

1. Plant variety (little is known of the varietal difference in the flavonoid level between Assam and Darjeeling varieties)
2. Blend of tea
3. Water : tea ratio
4. Duration of oxidation during processing
5. Brewing time (at a brewing time of 20 sec. the total flavonoids are 120 mg in a 235 ml brew, rising to 160 mg in a brewing time of 120 sec).

ANTIOXIDANT ACTIVITY OF TEA FLAVONOIDS

From the perspective of therapeutics, the antioxidant activity rather than the quantity of flavanoids, are crucial. Antioxidants protect the body against the damaging effects of free radicals produced naturally within the body. But overproduction of these free radicals due to environmental pollution, smoking or physiological disorders may disrupt the body's own antioxidant system resulting in the production of free radicals far in excess of what is good for health. An imbalance between free radical production and natural antioxidants could cause oxidative damages to proteins and DNA, the genetic material within the cells. A compound that prevents oxidative damage is, therefore, an antioxidant.

A plethora of evidence suggests strong antioxidant potentials of tea flavonoids in containing or suppressing the production of excess free radicals (Weisburger, 2000). Though most of the studies involved were laboratory-based and animal model-based (that mimic human system), they are nevertheless unequivocal that tea flavonoids can inhibit the actions of specified free radicals in human systems.

A few human studies are however emphatic that tea consumption reduces the oxidative damages to DNA in the human cells. Moreover, pilot studies using biomarkers established that green tea consumption for seven days could appreciably reduce DNA damage among both smokers and non-smokers (Lee et al., 1997). A balance of evidence indicates tea confers a mild beneficial effect on cardiovascular health, and even decreases the risk of heart diseases. As tea is the major source of flavonoids in tea drinking populations, it can provide a certain amount of heart health benefits as well (Banerjee, 1993).

TEA FLAVONOIDS AND CANCER

The evidence here is more gripping as a majority of research on tea therapeutics concentrate on cancer (Ganguly, 2003). Cancer occurs when the cells lose their natural control mechanism leading to their unregulated growth, eventually causing damage to the genetic information system coded in the DNA of cells. The damage process continues over a period of time leading ultimately to the cell growth rate in excess of that is needed for normal growth, repair and reproduction of cells.

The effects of tea (flavonoids) on cancer prevention are both strong and unique (Ganguly, 2003). Based on typical tea consumption by humans, it can be said that that both black and green tea have the potential to inhibit tumor formation in different parts of the body and are likely to be protective during the initiation, promotion and progression of both smoke and arsenic induced cancers. Tea flavonoids prevent initiation of cancer by inducing enzymes that help eliminate the toxins responsible for cancerous growth (Bhattacharayya et al., 2003). But there are other ways of tea's inhibitory actions (Table 11.5).

Table 11.5 How tea flavanoids inhibit cancer formation

1. Helping to maintain normal cell growth by blocking activation of an oncogene enzyme AP1.
2. Maintaining cell to cell communication.
3. Increasing apoptosis (programmed death of damaged cells).
4. Blocking blood supply to cancer cells which help in decreasing the spread of cancer cells.

CAVEATS

Tea, no doubt, holds promise and has the potential to prevent and cure serious human maladies, but it cannot be a panacea. Research is still needed, involving human subjects, as majority of the studies are laboratory centered. Epidemiological evidence concerning any human disease and their cure by tea is urgently needed from two types of studies:

1. *Case control studies* by which past tea drinking habits of people with newly diagnosed diseases are compared with people free from the diseases.
2. *Cohort studies* by which tea drinking habits of healthy people are linked to subsequent incidence of a disease after years of follow up.

Both these studies involve complex statistical designs to find out the risk or no risk parameters. But they will provide a realistic perspective of the therapeutic attributes of tea flavanoids at physiological levels.

KEY POINTS

The key points that emerge out of a corpus of research are:

1. Black or green tea permitting high flavonoid intake may help maintaining a healthy heart and reduce the risk of cancer. But this needs to be confirmed by population studies.
2. Tropical diseases need more attention. Little work, for example, is being done on the prevention or cure of gastric disorders using tea compounds.
3. Bio-availability of flavonoids. Focused research is needed to find out the degree and rate of absorption of flavonoids, as curative properties

of flavanoids are inexorably linked to their availability within the human system.

4. Do all tea varieties have the same level of curative or preventive properties, or some are more equal than others?

A SWOT ANALYSIS

An analysis of the current trend of research on tea therapeutics in India yields the following results:

Strengths

1. Mass database
2. Easily available resource
3. Involvement of high caliber pool of scientists
4. Availability of high pitched equipment and research infrastructure
5. Technical support–quality norms

Weaknesses

1. Over-dependence on laboratory studies
2. Sourcing of the biological materials used is mostly unknown
3. Lack of uniformity in the tea samples used
4. Little clinical research using human subjects

Opportunities

1. Efforts towards ameliorating tropical diseases using tea flavonoids
2. Improving nutrient values of tea
3. Relating and integrating flavour and quality with antioxidant properties of tea
4. Isolation of specific compounds of pharmacological importance with linkages to specific diseases
5. Identification of tea varieties with high antioxidant potentials
6. Emphasising the uniqueness of Indian tea as the source of powerful antioxidants and high quality

Threats

1. Competition from other beverages

2. Short term interest
3. Individualistic approach
4. Lack of coordinated research
5. Lack of intellectual property right on the special attributes of Indian tea in regard to its strong and unequivocal therapeutic properties

It is still premature to say tea indeed is a panacea and that all problems relating to its medicinal values have been solved (Banerjee, 1992). But it is correct to say that great progress has been made and more is in store. The progress made so far rests on the salient work as outlined above, and a sense of strategic policy is needed to fully explore the untapped potentials of tea.

REFERENCES

Banerjee, B. 1988. The cup that cheers. Nature, 322: 580.

Banerjee, B. 1992. Tea, health and unknowns. Assam Review, 81(5): 14-15.

Banerjee, B. 1993. Tea as a health drink. Co-ordination Committee of Planters Association, Kolkata.

Bhattacharyya, A., T. Chaudhury, S. Pal, S.K. Chattopadhay, G. Sa and T. Das. 2003. Apoptogenic effects of black tea on Ehrlich's ascetic carcinoma cell. Carcinogenesis. 24: 75-80.

Ganguly, D.K. 2003. Tea consumption on oxidative damage and cancer. Indian Council of Medical Research Bulletin, 3: 37-51.

Lee, I, Kim, Y. Hi. and M.H. Kang. 1997. Chemopreventive effects of green tea against cigarette smoke induced mutations in humans. J. Cell. Biochem., 27: 168.

Weisburger, J.H. 2000. Tea. In: The Cambridge World History of Food (eds: K. Kipple and K.C. Ornelas). Cambridge Univ. Press. Cambridge, pp. 712-720.

APPENDIX I
CHEMICAL COMPOSITION OF FRESH TEA FLUSH

	Molecular formula	*Molecular weight*	*% dry wt*
Flavanols			
(–) - Epicatechin	$C_{15}H_{14}O_6$	290	1–3
(–) - Epicatechin gallate	$C_{22}H_{18}O_{10}$	442	3–6
(–) - Epigallocatechin	$C_{15}H_{14}O_7$	306	3–6
(–) - Epigallocatechin gallate	$C_{22}H_{18}O_{11}$	458	9–13
(+) - Catechin	$C_{15}H_{14}O_6$	290	1-2
(+) - Gallocatechin	$C_{15}H_{14}O_7$	306	3-4
Flavonols and flavonol glucosides			
Quercetin	$C_{15}H_{10}O_7$	302	
Kaemperol	$C_{15}H_{10}O_6$	286	
Quercetin 3-rhamnoglocoside (rutin)	$C_{27}H_{30}O_{16}$	610	0.9–1.00
Kaempferol 3-rhamnoglucoside	$C_{27}H_{30}O_{15}$	594	
Quercetin 3-rhamnodiglucoside	$C_{33}H_{40}O_{21}$	772	
Kaempferol 3- rhamnodiglucoside	$C_{33}H_{40}O_{20}$	756	
Leucoanthocyanins acids and depsides			2–3
Theogallin	$C_{14}H_{15}O_{10}$	343	1
Ellagic acid	$C_{14}H_6O_8$	302	
Gallic acid	$C_7H_6O_5$	170	
Chlorogenic acid (4 isomers)	$C_{16}H_{18}O_9$	354	
p-Coumarylquinic (4 isomers)	$C_{16}H_{18}O_8$	338	
Total polyphenols			25–35
Caffeine	$C_8H_{10}N_4O_2$	194	3–4
Theanine	$C_7H_{14}N_2O_3$	174	1
Carbohydrates (simple)			3
Protein			15
Ash			5
Cellulose			7
Lignin			6
Pigments			0.5
Lipids			4

Frequently Asked Questions

What is the difference between normal leaf tea and herbal tea?

Herbal tea does not have even a leaf of *Camellia sinensis* – whose young shoots are used in making the standard black, green or oolong teas. Herbal tea is made from shrubs, not from tea plants, and generally lacks caffeine. Hence it is thought to have medicinal properties though evidence to support this contention is lacking. Some herbal teas are even suspected to be responsible for damage to liver.

Do black and green teas have similar effects?

Yes! Most studies show similar effects. The antioxidant properties of polyphenols from black and green teas destroy the reactive oxygen species (ROS) and prevent diseases in humans in identical manner. All teas, green, black or oolong have health promoting potentials. Even iced tea is beneficial from the perspective of health.

How much tea one might consume from health point of view?

Most research shows 5 to 10 cups daily might be optimal with a concentration of about 2% of tea, black or green, in the infusion.

Will addition of sugar affect the antioxidant effect of tea?

It might, but depending on the life style of the drinker, the effect would vary.

Can tea be the panacea for human diseases?

Not really! But the evidence of beneficial effects of tea consumption is gripping in some cases and suggestive in others as indicated below :

Disease	Evidence in humans
Cardiovascular	Strong
Cancer	Varying: population-wise
Diabetes	Suggestive
Obesity	Suggestive
Osteoporosis	Some
Arthritis	Suggestive
Neurological	Some
Bacterial (tooth decay)	Suggestive

(Siddiqui, Afaq and Mukthar: Int. J. Tea Sci. 2: 6-17: 2003)

What are antioxidants?

Antioxidants are compounds that help the body to fight harmful free radicals. The free radicals occur naturally in the body but excess amount increases the risk of coronary attacks and cancer.

How much antioxidants are present in a cup of brewed tea?

The amount may vary depending on the kind of manufactured tea, but it goes upto 200 mg per cup of standard size.

How to make antioxidants work?

Your daily cup by all means is one of the richest sources of antioxidants. Antioxidant activity of tea is greater than the antioxidant activity in 22 fruits and vegetables studied so far.

Can antioxidant catechins cause any risk if one drinks many cups of tea per day?

Huge amount of catechin intake over a lifetime will not harm. But it is better to be moderate. As long as the amount is less than 20 cups of tea per day there shouldn't be any problem. 7 to 15 cups per day gives 900 to 1500 mg of catechins per day. Even 1-2 gms of catachin per day will not cause any toxicity within the human system.

How antioxidants help?

Antioxidants, such as they are present in tea, are substances which scavenge excess free radicals generated within our body, and thereby prevent their damaging effects.

Are antioxidants found in tea natural?

Yes. Antioxidants occur naturally in tea.

Does adding milk to tea affect the antioxidants?

Adding milk (2%) to tea does not affect the antioxidant activity.

What is the level of caffeine in tea compared to other beverages?

Caffeine level in black tea is much lower than in other beverages. It is 40-50 ml per 130-ml cup, compared to 130-150 ml in a cup of coffee or 80-70 ml in some fizzy drinks.

What is oxidative damage?

The thing we need most is oxygen. Yet for inexplicable reasons, mitochondria – the finest microstructure in our body cells, misplaces the electrons, causing free radicals scatter around inside the cell, bonding indiscriminately with molecules. This mischief is called oxidation; it allows generation of the free radicals, which mangle DNA as well. This is the oxidative damage and the antioxidants in tea act against this damage.

What are free radicals?

Free radicals are reactive chemicals containing one or more unpaired electrons.

What is free radical reaction?

Free radicals are unstable molecules with nascent oxygen. To become stable they attack cell membrane, and may accept or donate one electron. But this process leads to generation of more free radicals *ad infinitum*.

How free radicals are generated in the body?

Free radicals are produced in the body naturally and as by-products of day to day metabolic activity. Deliberate production of free radicals ensures protection against foreign invaders in our body leading to disease state. But excess production of free radicals within the body due to pollution and other factors is harmful.

What are the damaging effects of free radicals?

Increased levels of free radicals damage cell membrane and disturb the delicate balance between cell structure and function. This harmful action is responsible for many human diseases affecting vital parts. These radicals work against body's immune system known as oxidative stress. This causes diseases like cancer.

Does our body have defense against free radicals?

The free radicals produced within the body are removed by body's own defense mechanism. But under diseased conditions, during aging or bad dietary intakes, body's defense mechanism becomes inadequate to remove excess free radicals generated.

How much medicinally important compounds are present in the dry black tea?

One gram of black tea contains 3.19 mg of epigallocatechin gallate and 9.50 mg of theaflavin.

Glossary

Antioxidants	Compounds that help the body fight harmful free radicals responsible for increased risk of coronary heart disease and cancer.
Apoptosis	Programmed death of cancerous cells, including DNA fragmentation. Also see p 110.
Atherosclerosis	A deposit of cholesterol leading to the narrowing of blood vessels and restriction of blood flow.
Caffeine	An alkaloid (1,3,7-trimethylxanthine) of purine group. It is a colourless compound with a bitter taste with a strong stimulating, both physical and mental, effect in human beings. Black tea contains about 9% caffeine.
Black tea	The most common form of tea popular world over. Making of black tea from two leaves and a bud involves a series of oxidations and condensations of certain compounds in the withered leaf by the rolling or CTC cuts. During the making of black tea no substances other than water is extracted (drying).
BTE	Black Tea Extract.
Cancer	Cancers are caused by progressive growth of the progeny of a single transformed cell, involving molecular alterations.
Carbohydrate	Refers to sugars, starch, cellulose etc., which are vital in the metabolism of all organisms.
Catechins	A group of organic substances, of which six have been identified in the tea leaf. They are derived from flavonols and are twenty times more powerful as antioxidants than Vitamin C. These are the compounds of great medicinal values.
CTC	Crush, tear and curling of tea leaves during the production of black tea. A specific machine does the job. The machine has two-toothed rollers which operate at different speeds

	in opposite directions to crush, tear and curl the leaves as they move continuously between the rollers.
Cell	The basic unit of life, the smallest living structure capable of functioning independently.
Chromosomes	Tightly coiled strands of genes (DNA) located in the nucleus of every eukaryotic cell. Each chromosome has a fixed number of genes, and every species has a characteristic number of chromosome pairs – humans have 23 pairs, mice have 19 and pea plants have 7.
Clone	An exact genetic replica, of one specific gene or of an entire organism.
Cloning	The process of a sexually producing an identical copy of a parent organism, be it a cell or, say, a cutting from a tree. Bacteria, which reproduce by simply splitting in half, always produce clones.
Culture	A particular strain or kind of organism growing in a laboratory medium. Also called a cell culture.
Cytoplasm	The living stuff inside all cells, but outside eukaryotic cells nuclei.
Deoxyribonucleic Acid (DNA)	The chemical chains that contain the genetic information in all organisms (with the exception of a small number of viruses in which the hereditary material is ribonucleic acid – RNA). The information coded by DNA determines the structure and function of an organism.
Drying	A stage in the processing of tea by which the fermented (oxidized) tea is dried to get the finished product which is stable. Drying is done in tea dryers, reducing the moisture level to 3%.
(–) EC (epicatechin)	Polyphenolic compound in tea leaf that undergoes changes during fermentation i.e. oxidation.
(–) ECG (epicatechin gallate)	Also a polyphenolic compound in the tea leaf that is oxidized during fermentation.
(–) EGC (epipgallocatechin)	Polyphenolic compound that is oxidized during fermentation.
(–) EGCG (epipgallocatechin gallate)	This is also a polyphenolic compound that undergoes oxidation during fermentation during the processing of tea. One gm of black tea contains 9.50 mg of epipgallocatechin gallate.
Enzymes	Proteins that control the various steps in all chemical reactions – usually speeding them along.
Fermentation	Fermentation, misnomer, is in reality an oxidative process during the manufacturing of black tea. It involves a series of enzymatic reactions, the crucial being the oxidation of polyphenols with the help of the enzyme, polyphenol oxidase. These reactions result in the formation of two

	groups of colouring compounds – the yellow group (theaflavins) and the red group (thearubigins).
Flavonoids	These are a group of polyphenolic compounds which have strong antioxidant effects. One cup of tea supplies 200 mg of flavonoids. Flavonoids include tannins, catechins, theaflavins and thearubigins all which are of medicinal importance.
Flavonols	This is a sub-group of flavonoids with highest antioxidant property.
Gene	A segment of DNA specifying a unit of genetic information: an ordered sequence of nucleotide base pairs that produce a specific product or have an assigned function.
Genome	The complete complement of genetic material in any individual.
Genotype	The particular set of genes present in an organism's cells: organism's complete genetic recipe. Compare phenotype.
Green tea	A variety (like black tea) of made tea. In making the green tea fermentation i.e. oxidation, of the leaf is totally avoided and is least pressed. Slow process of withering (of black tea) is replaced by the more rapid steaming process. The enzyme deactivation during the manufacturing of green tea ensures that the primary polyphenols as much are passed into the finished product.
Host	An organism harbouring another organism.
Hybrid	The offspring of genetically dissimilar parents, whether cells, plants, or animals.
in vitro literally, "in glass"	Describes biological reactions taking place in test tubes or any laboratory containers – all of which are artificial systems.
in vivo literally, "in life"	Describes biological reactions taking place inside the living organisms.
Mitochondria	Microstructures within the cells, they convert chemical energy into more readily useable high energy phosphate bonds (of ATP). They are also the major intracellular sources of ROS production.
Mutagen	An agent or process that can causes mutations, such as radiation or chemicals. Also see p. 111.
Mutagenesis	The process of mutation in the genetic material of an organism mutant. A strain differing from its parent because of mutation.
Metastasis	Translocation of cells by the formation of new blood vessels: common in cancerous situations. Also see p. 110.
Nucleic acids	There are two nucleic acids: DNA and RNA, made up of long chains of molecules called nucleotides.

Nucleotides	The building blocks of DNA and RNA. Each nucleotide molecule is composed of phosphate, sugar and one of four bases. These bases form condons, which when strung together form genes, which in turn link together to form chromosomes.
Nucleus	The central compartment in all eukaryotic cells that houses most of the DNA in higher organisms.
Pathogen	Any organism capable of producing disease.
Phenotype	The observable characteristics of an organism. The phenotype results from the interaction between the genetic constitution (genotype) of an organism and its environment.
Plasmid	Free-floating circular ring of DNA that can be integrated into genes and chromosomes. Plasmids can ferry genes and thus are favourite tools of transformation for genetic engineers.
Recombinant	A cell or clone of cells resulting from recombination of genes.
Recombinant DNA (Abbrr. rDNA)	As a product: Fragments of DNA from two different organisms, such as a human and a bacterium, spliced together in the laboratory into a single molecule. As a process: A broad range of techniques involving manipulation of the genetic material of organisms; "genetic engineering".
Oesterogen	Female reproductive hormone.
Oolong tea	Semifermented (semioxidised) tea: primary polyphenols are therefore partly oxidized. It is intermediate in character between black and green tea. It has good medicinal property.
Oophorectomized	Removal of ovaries, often done during experimentation.
Polyphenols	A large group of compounds of therapeutic importance forming about 30% of the dry matter of tea shoot. The group includes flavanol or catechin and its gallates i.e. gallocatechin, epigallocatechin, epigallocatechin gallate and epicatechin.
Polyphenol oxidase	An enzyme responsible for the oxidating of polyphenols during fermentation (oxidation) in black tea manufacture. It is localized in the chloroplasts and within the epidermal cells.
Rolling	A stage in orthodox black tea processing which distorts the semipermeable membranes of leaf cells. This enables mixing of polyphenols with polyphenol oxidase in the presence of oxygen.
ROS	Reactive Oxygen Species.
Species	The taxonomic group of like-individuals. Variants within species are called strains.

Theaflavins	Yellow compounds formed during the fermentation of tea. They add to the quality, briskness and brightness of tea, apart from being of considerable therapeutic value.
Thearubigins	These are red compounds formed during the fermentation process of tea. They are formed by the condensation of theaflavins. Thearubigins are also of therapeutic importance and generally add to the strength and odour of tea.
Virus	A small genetic element, composed DNA or RNA and protected by a protein coat, that is able to alternate between intracellular and extracellular states. Viruses do not have metabolic functions and cannot reproduce without first infecting and then taking over a living cell.
White tea	Tea manufactured from buds (and flowers) of tea by avoiding oxidation.
Withering	The first stage in tea manufacturing causes a reduction in the moisture level of the green leaves. Some biochemical changes also occur during withering.

Index

About the Editors

Dr. B. Banerjee, Advisor to the National Tea Research Foundation, India, had his education at Illinois, London and Oxford, was Nebuta Fellow at the International Agricultural Centre, Netherlands. He manned top research positions at Tea Research Association – Tocklai Experimental Station, including heading it. Dr. Banerjee held the World Bank Chair in Agricultural Faculty at the University of Nairobi, concurrently at Washington, was correspondent for the Special Committee on the Problems of Environment of UNESCO and US based International Panel Studying the Impact of Climatic Fluctuations on Crop Yields.

Dr. Banerjee authored six books including Tea Production and Processing (Oxford & IBH), is contributory author to Ecosystem of the World (Elsevier), Tea Cultivation to Consumption (Chapman and Hall), Tetranychids (Elsevier), Plantation Crops (Oxford & IBH) etc., has some 80 research papers in International Journals of Professional repute.

Dr. Banerjee is an awardee of BCCI, Bharat Chamber of Commerce and Industry, S N Bangur Foundation, The Ecology Society USA, The British Council, Commonwealth Bureau etc. He is associated with the National Academy of Sciences.

Dr. T.C. Chaudhuri is Director of Research in the Tea Board of India. His earlier assignments include scientist in ICAR and silk research in the Central Silk Board. He is proactive in the planning and coordination of tea research in India, including biodiversity and biotechnology projects. He is a member of the Councils of Management of Tea Research Association, IHBT and IIPM. He is Secretary of the National Tea Research Foundation (NTRF).

Dr. Chaudhuri is a member of the Working Group under Inter Governmental Group, who monitor and regulate international collaboration and coordination of chemical residues. He is a member of National Committees on CODEX, International Standards Organisation (ISO) and the Central Committee of Food Standards in the PFA, and committees of BIS for quality and packaging of tea. Dr. Chaudhuri is involved with standardization of quality parameters for tea.